Topics in Recreational Mathematics 2/2015

Editor-in-chief

Charles Ashbacher
5530 Kacena Ave
Marion, IA 52302 USA

cashbacher@yahoo.com

Assistant Editor

Rachel Pollari

Artwork

Caytie Ribble

Alphametics Contributor

Paul Boymel

Problem Contributor

Lamarr Widmer

Technical Advisor

Gisela Hausmann

ISBN-13: 978-1508617099

Contents

INTRODUCTION

This is the second book designed to continue the legacy of publishing recreational mathematics that was started by Joseph S. (Joe) Madachy. Joe passed in 2014 so the opening item is a tribute to his work in developing and sustaining a publishing outlet for recreational mathematics. A few other smaller items of his work, from both **Recreational Mathematics Magazine** and **Journal of Recreational Mathematics**, also appear in other sections.

I would like to thank Sage Publications, the current owners of the rights to **Journal of Recreational Mathematics (JRM)**, for giving me permission to publish the solutions to the problems that appeared in issues 37(3) and 37(4) of **JRM**. Since these solutions were scheduled to appear in issues 38(3) and 38(4) of **JRM** and the final issue was 38(2), they otherwise would not have been published. The solutions to 37(3) problems appear in this book and the solutions to the 38(4) problems will appear in the third book that I have already committed to. When that book appears, nearly all of the material left from the termination of **JRM** will have been published.

Paul M. Sommers was the most prolific contributor to **JRM** when I was editing the manuscript content and the two papers of his that are in this book were sent to me in the hopes they would be published in **JRM**. At this time, I am performing a deep mining operation: looking through items that were submitted to **JRM** but never published and re-evaluating them. The slow rate of publication of **JRM** in its last years meant I had to be very selective in what I accepted.

One tactic I adopted was a reluctance to publish longer pieces. If a longer one appeared, readers that didn't like it were less likely to read it. Using two shorter pieces instead meant that it was more likely an arbitrary reader would find something that interested them.

I am actively soliciting material for what could become the fourth book in this series. If you have anything in the area of recreational mathematics that you are interested in having published, pass it on to me. I only accept electronic copies and now require an abstract with all submitted papers. If you are not sure if your work is recreational mathematics, understand that it is a very broad field open to interpretation. If you have any doubts, send me an email asking the appropriate questions.

Hope you enjoy the material you will find here!

I encourage all readers to send their comments to me at cashbacher@yahoo.com

Charles Ashbacher

TRIBUTE TO JOSEPH S. MADACHY

Charles Ashbacher

In March of 2014, Joseph (Joe) S. Madachy passed on. He was a driving force in the area of recreational mathematics, turning it from a field of occasional publication in various and scattered books, periodicals and journals to one with a regularly published venue. His image appears in figure 1.

Figure 1

In February of 1961 Joe edited and published the first issue of **Recreational Mathematics Magazine (RMM)**, the first journal devoted exclusively to the fun side of mathematics. The cover of that issue appears in figure 2.

The style of the journal would be set in that first issue. The table of contents (TOC) is displayed in figure 3. There are several things to note in the TOC. First, Joe actively solicited material and comments from readers. Over the years, while I contributed to **JRM** and before I succeeded Joe as manuscript editor, he was regularly in touch with me about my work, offering suggestions for improvement as well as ideas for new avenues of exploration.

Three entries in the TOC of the first issue of RMM - alphametics, puzzles and problems and book reviews - appeared as regular features in all subsequent publications. When **Journal of Recreational Mathematics** ceased publication with issue 38(2) in December of 2014, these areas were still featured in every issue.

Unfortunately, **Recreational Mathematics Magazine** had a short life, ceasing publication in 1964 after 14 issues.

Figure 2

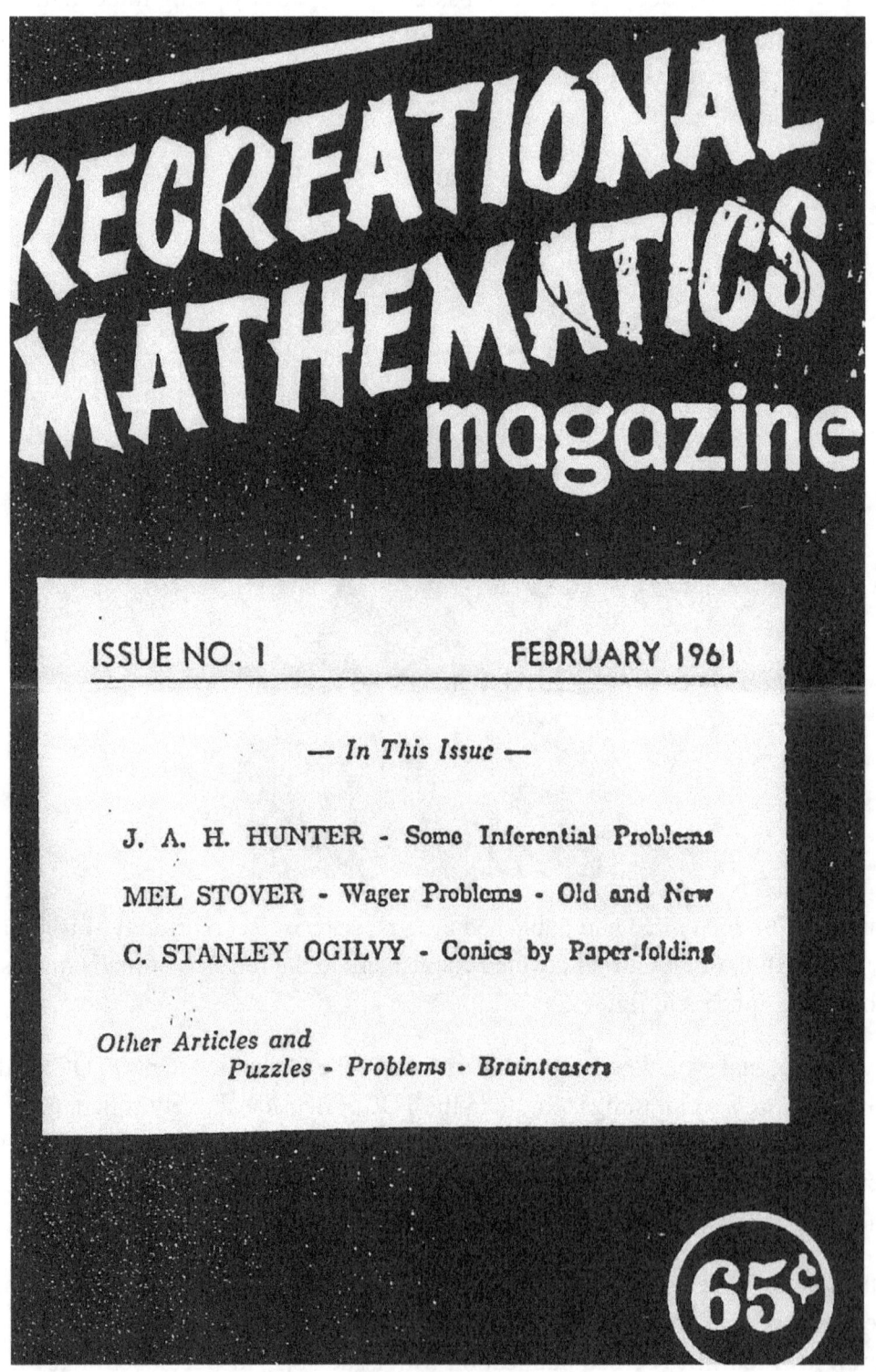

Figure 3

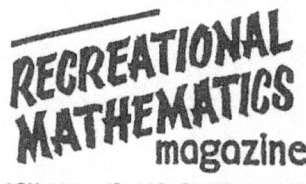

RECREATIONAL MATHEMATICS magazine FEBRUARY 1961 ISSUE NUMBER 1

BOX 1876 IDAHO FALLS, IDAHO PUBLISHED AND EDITED BY JOSEPH S. MADACHY

Contents

RECREATIONAL MATHEMATICS MAGAZINE published bimonthly by Joseph S. Madachy at The Falls Printing Co., Idaho Falls, Idaho. Application to mail at second class postage rates is pending at Idaho Falls, Idaho. Subscription rates: $3.00 per year for teachers, students and libraries; $3.50 per year for the general public; these rates are worldwide. Reprints of any material must be requested within the month of publication. reprint costs are 5¢ per page. All correspondence concerning changes of address, subscriptions, reprints and manuscripts should be sent to The Editor, Recreational Mathematics Magazine. Box 1876, Idaho Falls, Idaho.

The following is the letter from the editor that Joe wrote as the opening piece of **Recreational Mathematics Magazine**.

From the Editor

For those many people who placed their subscription orders long before **Recreational Mathematics Magazine (RMM)** *was scheduled to appear the receipt of this issue must be accompanied by some such thought as "It's about time!" The editor is in complete accord with that same thought. The task has been long and hard, but deeply satisfying.*

RMM *is a magazine intended to fulfill the desire of many for a unique periodical entirely devoted to the strictly lighter side of mathematics. Here you will find no advanced calculus or number theory requiring a Ph. D. in mathematics. But you will find such things as number curiosities and tricks, paper-folding creations, chess and checker brainteasers, articles about mathematics and mathematicians, discussions of certain aspects of higher mathematics and their application to everyday life and to puzzle solving. You will find word games, geometric dissections, magic squares, map-coloring problems, cryptography, and many other topics generally included in the fields of puzzles and recreational mathematics.*

It is hoped that some of the effects of reading **RMM** *is the stimulation of an interest in mathematics in general, the acquiring of a deeper appreciation of the values and beauties of mathematics and the promotion of at least a certain degree of logical thinking. It is hoped, also, that both teachers and students will find material aplenty to augment formal mathematical education.*

RMM *could not have reached its present stage without the help of many people. At the head of the list are the many subscribers who expressed confidence and hope enough to subscribe before seeing even a review of the magazine itself. Thanks and deep appreciation are also extended to: Martin Gardner, who conducts the "Mathematical Games" section in* **Scientific American** *and who has spread the word of* **RMM** *throughout the world and given me aid beyond measure; to Jim Hunter of Toronto who contributed the first article in this issue and who saw to it that many of his fellow Canadians were contacted; to Dr. H. V. Gosling of Kingston, Ontario, whose material has been most welcome and whose many letters gave much-appreciated encouragement and moral support; to Bob Underwood of the University of North Carolina who suggested, among other things, the idea of a Readers' Research Department and to many, many others who must remain nameless due to lack of space.*

It seems to be the custom in dedicatory editorials and prefaces to extend unending accolades to the one person most closely allied to the editor – his wife. I cannot break custom. Without her continued ego-bolstering encouragement, her hours of monotonous typing and the tolerance *extended to all puzzle nuts, it is doubtful that **RMM** could have come into existence this soon, if at all.*

1 February, 1961   *J. S. M.*

One thing that is very clear from this letter - to which I can personally attest - is that Joe was very generous in giving encouragement, publicly thanking the people that assisted him and allowing others to take credit. Not once in my correspondence with him did he ever raise an issue about receiving credit for an idea or a contribution. Joe received credit for many items, but he legitimately could have claimed joint authorship for many more.

In 1967, the dormant fire of regular publications in recreational mathematics was rekindled when Greenwood Press asked Joe to start a journal in recreational mathematics, this time called **Journal of Recreational Mathematics (JRM)**. The beginning of **JRM** was much like that of **RMM**, where it contained similar types of material. Joe once again set the tone for the publication in his opening letter where he praised all who helped him in the restart. The front cover of the first issue of **JRM** appears in figure 4.

The table of contents of the first issue of **JRM** appears in figure 5. Once again, the reader can see the fundamentals of the content established with sections containing problems, book reviews and alphametics.

Joe introduced the newly minted **JRM** in his usual style, by setting down his goals and passing out credit to others rather than taking all of it himself. The original letter from the editor that appeared as the first item in the first issue of **JRM** is reproduced here.

From the Editor

*There is no doubt about the popularity of mathematical recreations. Martin Gardner's three books based on his column in **Scientific American** have been bestselling items in the field; dozens of books devoted to recreational mathematics and puzzles grace the monthly lists of new books; the professional mathematics journals' most popular features are the "Problems" sections, and these problems are seldom of the textbook variety – they include geometric diversions, number games and tricks, and alphametics. Really, it is unnecessary to go on – readers of this journal already know.*

*The birth of the **Journal of Recreational Mathematics** with this first issue may remind many of its readers of previous periodicals devoted to mathematical recreations: **Sphinx**, edited in the late 1930's by Maurice Kraitchik, and **Recreational Mathematics Magazine**, edited by myself from 1961 – 1964. I received many letters from persons familiar with one or both of these past publications, and all are of the same mind as to the general value of recreational mathematics in stimulating an interest in mathematics and in mathematical education.*

Figure 4

Volume 1   Number 1   *January, 1968*

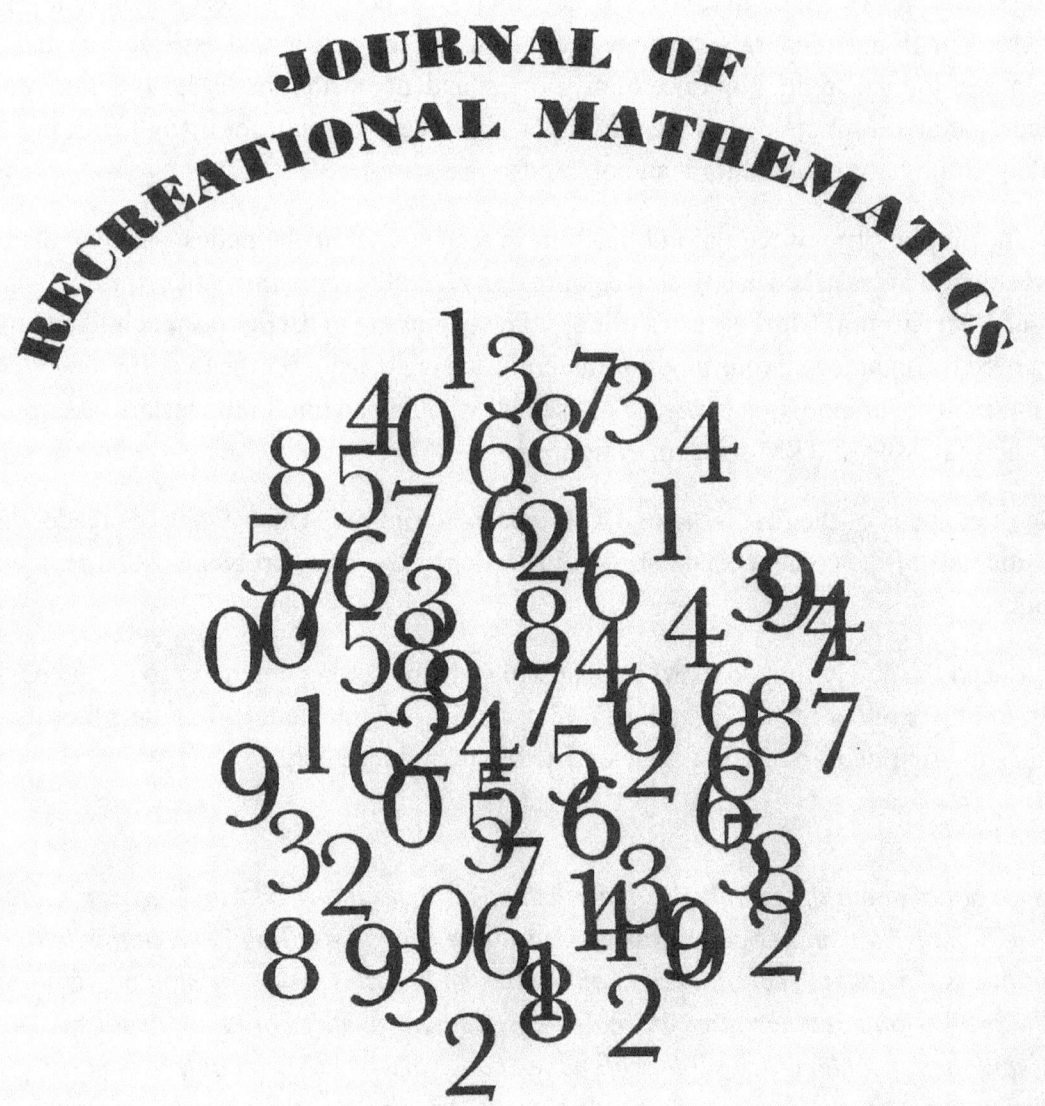

Published by Greenwood Periodicals, Inc.

Figure 5

JOURNAL OF
RECREATIONAL MATHEMATICS

Volume 1
Number 1
January, 1968

Contents   1

---

*The variety of material possible in this field is only hinted at in this first issue. A list of some of the articles to appear in **JRM** can be seen on page 44, and I will add just a few more: "Approximate Geometric Dissections" by Harry Lindgren; "Various Curiosa on 1968;" "Compound Games With Counters" by Cedric A. B. Smith; "The Mersenne Primes and Perfect Numbers" by Sidney Kravitz, et al.; and more problems and book reviews.*

11

*I hope to expand the Editorial Board, increase the number of referees (volunteers?), publish brief biographies of authors and Board members in future issues, and, in general, continuously strive for a bigger and better **JRM**.*

*All issues of any publication show birth pangs. **The Journal of Recreational Mathematics** is no exception. Greenwood Periodicals, Inc. has worked hard at putting into print the articles and material submitted. The editor, however, assumes responsibility for the choice of material. We all hope that slight imperfections will be pointed out, but that they will not be used to condemn us. Criticism will be welcome, as well as commendation.*

*I wish to extend my personal thanks to the many old friends who remembered **RMM** and offered aid during the past months. I do not have sufficient space to fully extend my appreciation of the efforts exerted by the staff at Greenwood Periodicals Inc. Although there is not enough space to list all those who helped, I cannot conclude without naming at least a few of those individuals who graciously extended assistance: Martin Gardner, Vern E. Hoggatt, Jr., J. A. H. Hunter, Howard C. Saar and Charles W. Trigg.*

January 1968   *J. S. M.*

In 1982, Joe edited **Ten Year Cumulative Index to the Journal of Recreational Mathematics** (ISBN 0895030209) that was published by Baywood Publishing. At the start of the book, he wrote a personal history of his involvement in recreational mathematics and it is reprinted here. His honesty in admitting his experience of a bankruptcy while trying to make **Recreational Mathematics Magazine** a viable entity demonstrates how passionately he believed in the importance of recreational mathematics in the world.

RECREATIONAL MATHEMATICS:

A Personal History

Joseph S. Madachy

*   **Recreational Mathematics Magazine** *had its beginning in June, 1960, when I conceived the idea of a journal entirely devoted to recreational mathematics and containing material that would appeal to the informed layman. The content, therefore, would avoid treatment that would require a college degree for its understanding. Through my interest in recreational mathematics, I was aware of earlier defunct publications and the existence of magazines that carried columns devoted to puzzles and games. There were also two or three foreign efforts, containing only a few pages each, devoted to the subject, but there appeared to be no English-language publication dealing entirely with recreational math.*

*I wrote to Martin Gardner, the editor of the "Mathematical Games" column in **Scientific American**, explaining my idea. Gardner was enthusiastic and encouraged me to proceed; even supplying a list of probable contributors. The list proved doubly beneficial; it supplied enough publishable material to fill a few issues plus other names of possible contributors. It became very obvious that there was a plentiful resource of unpublished material available.*

*Totally committed to my one-man show of Editor, Publisher and Publicity Manager, my title **Recreational Mathematics Magazine** determined; my format planned; I generally wore myself out promoting the whole idea, with direct mail and magazine ads.*

*In early February, a short eight months after the original concept, the first issue of **RMM** was mailed to about 1100 subscribers. By December 1961, six issues later, the number of subscribers had increased to over 5000. Everything seemed to be going great! Several major business errors changed the picture, as well as the fact that a near one-man-show for an undertaking of this scope was just too demanding. My load was lightened considerably with the help of several faithful readers and contributors who became associate or assistant editors, but I had requested help too late. After fourteen published issues, I declared bankruptcy in 1964. The **RMM** fiasco was chalked up as a great experience!*

*In mid 1967, Greenwood Press contacted me (at Martin Gardner's suggestion) with an offer to become Editor of a new recreational journal they were preparing to publish. After ten seconds of careful thought, I accepted. The new journal, **Journal of Recreational Mathematics**, - after much familiar planning and work – was distributed in early 1968. Armed with the knowledge that a journal of this nature could not achieve its basic goal of presenting mathematics in a unique manner, without help, an Editorial Board was immediately formed. My "staff" on the old **RMM** had consisted of Associate Editor J. A. H. Hunter, with assists from Howard C. Saar, Dmitri E. Thoro, and Harry L. Clawson. The new **JRM** started with an editorial board that included Hunter and Saar, and additionally Leo Moser.*

*As further issues of **JRM** were published, new members were added to the Board. From a three-member start, the Board grew to its present nine-member size. The help of the following Board members, during their short or long terms, was and still is incalculable: David L. Silverman, David A. Klarner, Benjamin L. Schwartz, Charles W. Trigg, James B. Haley, Jr., Harry L. Nelson (who became the Editor in 1976), Steven Kahan, myself (when Nelson became Editor), Harvey Hindin, Friend H. Kierstead, Jr., John Brinn, Romae Cormier, and Frank Rubin.*

*No doubt exists in my mind that the help of all these Board members, the cooperation and extreme patience of both publishers (Greenwood Press published the first five years of **JRM**, Baywood Publishing Company, since then), and, most important of all, the legion of faithful readers and steady contributors, have given **JRM** a unique staying power. It has survived the already mentioned changes of editors and publishers.*

Special mention must be made of Harry Nelson's connection. For nearly twenty years he has been a reader, problem and puzzle solver par excellence, major contributor, assistant editor, Editorial Board member, or Executive Editor of both **RMM** *and* **JRM***.*

The **Journal of Recreational Mathematics** *is not the same as the old* **Recreational Mathematics Magazine***.* **RMM** *was a youngster. Entering its fourteenth year,* **JRM** *is a testament to my original premise: there are many people out there who want and enjoy this kind of material.*

Joe edited the **Journal of Recreational Mathematics** for thirty years, at which point I had the honor of succeeding him as the editor in charge of manuscripts. He was a giant in the field, second only to Martin Gardner in his contributions to recreational mathematics. I have always tried to perform at that high level, not always succeeding. Although he and I never met, we exchanged a lot of messages. I still vividly remember the unexpected phone call when he asked me to join the editorial board of **JRM**. I felt honored and flattered and was even more so when he called me years later and asked me to assume the editorship. Like so many others, I owe him a great deal. He will be missed.

MATHEMATICAL CARTOONS

drawn by Caytie Ribble

Thoughts of a Setistic Person

$\emptyset$ $\{1\}$
R^2 Z N $\{1,2,4,8...\}$
$\{2,3,5,7,11,13,17,19...\}$ $\aleph_0$
ω_1

A=B if and only if
A$\subseteq$B and B$\subseteq$A

A posteriori reasoning

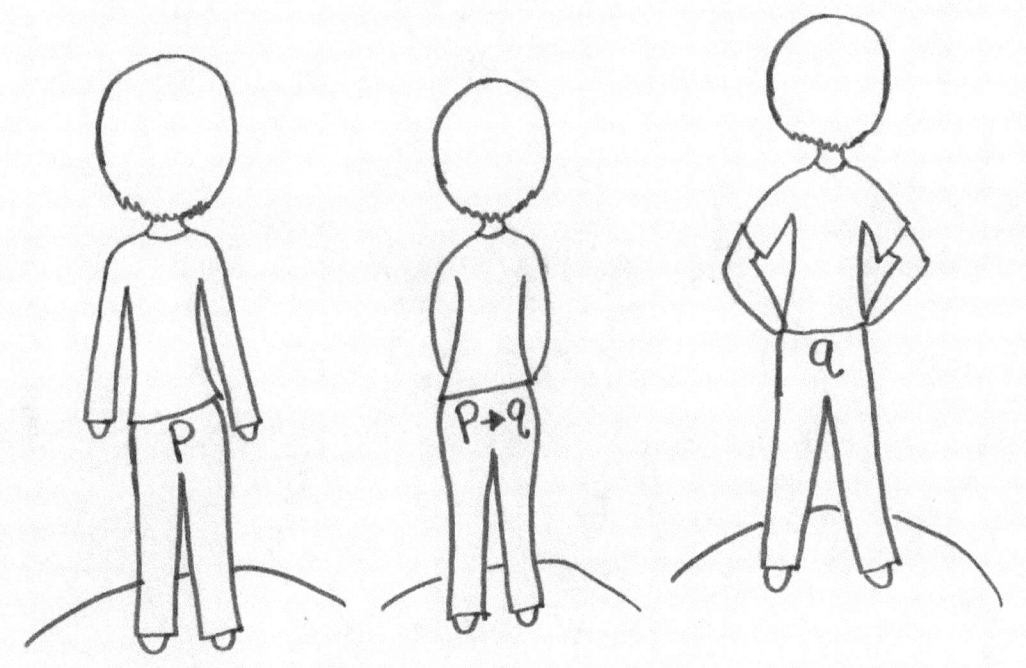

HERE'S THE SCOOP: GROUND BALLS WIN LACROSSE GAMES

Peter R. Smith

Russell K. Banker

Paul M. Sommers

Department of Economics

Middlebury College

Middlebury, Vermont 05753

psommers@middlebury.edu

Abstract

A key metric in lacrosse is the number of times a team scoops up a ground ball. Box scores of every men's lacrosse game for Division III colleges in the New England Small College Athletic Conference (NESCAC) between 2005 and 2009 were examined to gauge the importance of ground balls to winning. A series of chi-square tests, one for each school in NESCAC, shows that ground balls were critical to success. For each school, the authors also use regression analysis to show a strong direct relationship between the team's margin of victory and the percentage of ground balls won.

Lacrosse, a game popular with Native Americans long before Europeans first settled in North America,[1] has received scant attention from statisticians relative to other "stick-and-ball" games, like baseball and golf. Easily quantifiable measures of performance abound in baseball (for examples, slugging average, on-base percentage, and earned run averages) and golf (for examples, distance of a player's tee shot, greens in regulation – or the percentage of time that a player gets his ball onto the putting green within two shots of par, and the average number of putts per hole). But, that is not the case in lacrosse.

In this brief note, we focus on one key statistic: ground balls. In lacrosse, a "ground ball" is when the ball is scooped up off the ground. When the ball is on the ground, however, neither team has possession. And, since records are kept of the number of times in a game each team successfully scoops up a ground ball, we will determine whether or not "ground balls" are critical to team winning.

All colleges in the New England Small College Athletic Conference (hereafter NESCAC) play Division III lacrosse. We examined the box scores of every men's lacrosse game for all but one of the ten NESCAC schools from 2005 through 2009.[2] Over the five-year period, Wesleyan and Williams won one NESCAC title each (in 2009 and 2008, respectively). Middlebury College won seven consecutive NESCAC championship games since the inception of the NESCAC Men's Lacrosse Tournament in 2001. Of the ten NESCAC schools examined here, Middlebury College had the best overall record since 2005 (69 wins and 19 losses for a .784 winning percentage) and Colby had the worst (26 – 44, .371).

To test the null hypothesis that ground balls won is not related to games won, we use a chi-square test. All lacrosse games are either won or lost; there are no ties, with as many "sudden-death" overtime periods as are necessary to break ties at the end of regulation. We excluded games where the number of ground balls for and against each team was the same.[3] All games for each NESCAC school are divided into four groups as shown, for example, by one NESCAC school, Middlebury College, in Table 1. In 54 [9] games, Middlebury College recorded more [fewer] ground balls than her opponent (that is, "ground balls for" exceeded [was less than] "ground balls against") *and* Middlebury College won [lost] the game. The calculated chi-square (χ^2) statistic is 5.476. The probability that the chi-square test statistic will be as large as this (or larger) is only .019. The diagonal elements in the contingency table in Table 1 were much larger than would be expected if the null hypothesis were true. That is, when Middlebury College scooped up more (fewer) ground balls than her opponent, more often than not Middlebury won (lost) the game. The results for all of the other NESCAC schools are reported in Table 2. And, in every case, the diagonal elements of the contingency table are disproportionately large. Still, in other words, ground balls (for) are indeed key to success.

 Table 3 shows the results of regressing (for each game at each NESCAC school over the five-year period, 2005-09) the margin of victory (that is, "goals for" minus "goals against") against the percentage of ground balls won. How well the regression line fits the scatter of points (as measured by the R^2 or the coefficient of determination) is best for Amherst and is shown in Figure 1. In every case (that is, for all ten NESCAC schools), there is evidence of a strong direct relationship between the team's margin of victory and the percentage of ground balls won. One can use the regression results in Table 3 to find the percentage of ground balls won above which the winning margin is greater than or equal to "1" goal (as reported in the last column of Table 3).

Three of the four schools with the best winning records over the five-year period [Middlebury, .784 winning percentage; Wesleyan, .766; and Tufts, .679] had minimum threshold percentages of ground balls won less than 50 percent of the time. For these schools, while "ground balls for" is important, other facets of the game (collectively) might be even more important (for examples, goalie save percentage, extra-man opportunities, and man-down defense). Curiously, the school with the worst winning record over the five-year period [Colby, .371] needed to win, on average, at least 58.4 percent of ground balls to win a game.

Concluding Remarks

The evidence presented here for Division III schools in the NESCAC conference suggests that ground balls win games in men's lacrosse. Lacrosse involves numerous body and stick checks (as long as contact is from the front or side and above the knees but not above the shoulders and one's opponent has possession of the ball or is within five yards of a loose ball). As a consequence, the ball is knocked loose from the pocket of a player's stick and frequently ends up on the ground. An errant pass or shot suffers a similar fate. What we have found is that the team that wins the battle for ground balls more often than not wins the game. After all, more ground balls scooped up leads to more possessions. More possessions in turn lead to more scoring opportunities. And, more possessions by one team decreases the opposing team's time of possession and scoring chances.

Table 1

Contingency Table Relating

Ground Balls Won and Games Won,

Middlebury College, 2005-2009

	Ground Balls Won?	
	Yes	No
Won Game?		
Yes	54	14
No	10	9

Table 2

Summary of Chi-Square Tests,

Men's Lacrosse in NESCAC,

2005-2009

School	Won Game, Won GBs	Won Game, Lost GBs	Lost Game, Won GBs	Lost Game, Lost GBs	χ^2_{Calc}	p-value
Amherst	24	5	9	25	19.880	<.001
Bates	17	9	8	25	10.081	.001
Bowdoin	33	10	13	16	7.647	.006
Colby	21	2	17	24	15.174	<.001
Middlebury	54	14	10	9	5.476	.019
Trinity	30	3	13	23	22.017	<.001
Tufts	40	13	8	16	12.494	<.001

Table 2

Summary of Chi-Square Tests, Men's Lacrosse in NESCAC,

2005-2009 (continued)

Wesleyan	51	21	1	21	29.957	<.001
Williams	21	9	5	23	15.921	<.001

Table 3

Summary of Regression Results, Men's Lacrosse in NESCAC,

2005-2009

Dependent variable: Winning margin (Goals for – Goals against)

School	Constant	Percentage of ground balls won	R^2	Percentage of ground balls won, winning margin ≥ 1
Amherst	-24.702 (<.001)[*]	.482 (<.001)	.482	53.3
Bates	-15.454 (<.001)	.313 (<.001)	.328	52.6
Bowdoin	-13.152 (<.001)	.262 (<.001)	.257	54.0
Colby	-22.770 (<.001)	.407 (<.001)	.358	58.4

Table 3

Summary of Regression Results, Men's Lacrosse in NESCAC,

2005-2009 (continued)

Middlebury	**-14.443**	**.326**	**.209**	**47.4**
	(<.001)	**(<.001)**		
Trinity	**-21.016**	**.404**	**.401**	**54.5**
	(<.001)	**(<.001)**		
Tufts	**-15.947**	**.341**	**.376**	**49.7**
	(<.001)	**(<.001)**		
Wesleyan	**-14.798**	**.352**	**.315**	**44.9**
	(<.001)	**(<.001)**		
Williams	**-11.644**	**.260**	**.282**	**48.6**
	(<.001)	**(<.001)**		

[*]Numbers in parentheses are p-values.

Figure 1

Margin of Victory *v.* Percentage of Ground Balls Won,

Amherst College, 2005-2009

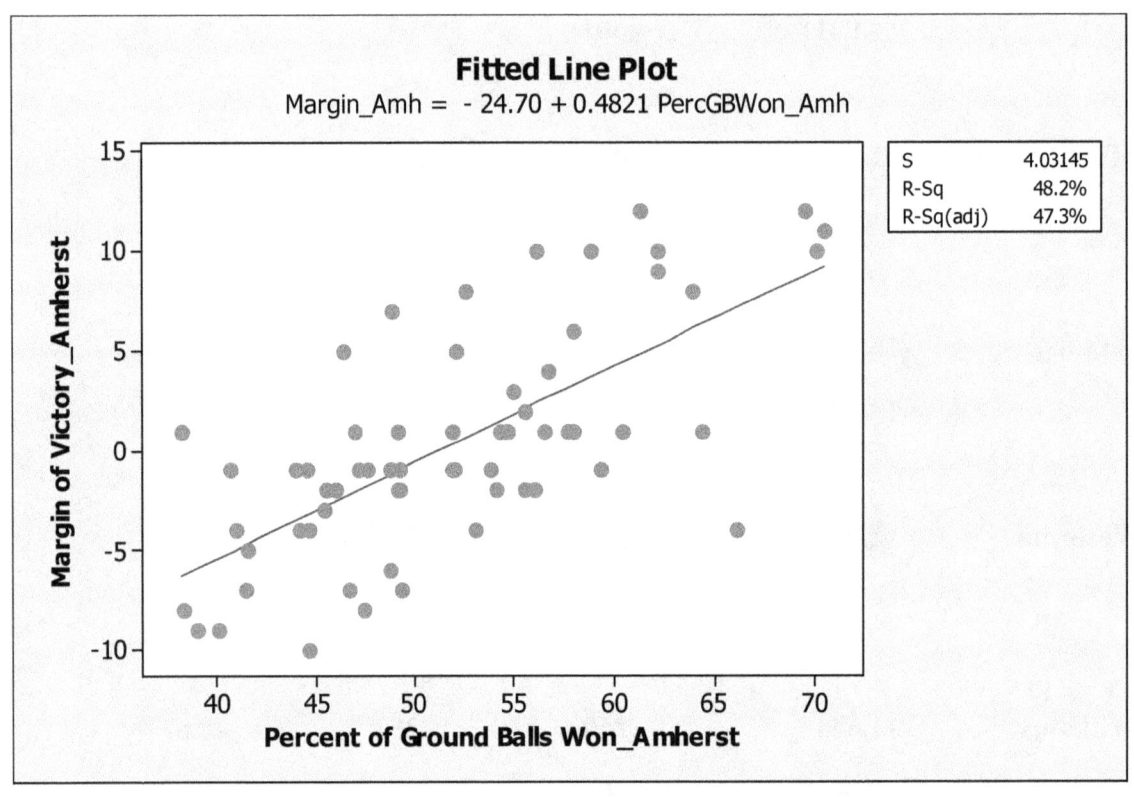

References

1. D.M. Fisher, **Lacrosse: A History of the Game**, Johns Hopkins University Press, Baltimore, MD, 2002.

2. D.G. Pietramala and N.A. Grauer, **Lacrosse: Technique and Tradition**, Johns Hopkins University Press, Baltimore, MD, 2006.

AN EXAMPLE OF A SMARANDACHE GEOMETRY

Professor Ion Patrascu

National College Fraţii Buzeşti

Craiova, Romania

Abstract

For centuries it was thought that the geometry codified by Euclid and based on the parallel postulate

Given any line l and a point p not on l, one and only one line can be drawn through p parallel to l.

was the only geometry that existed. This idea was overturned when two other geometries based on a different number of parallel lines being drawn through p were discovered. In a hyperbolic geometry it is possible to construct infinitely many lines parallel to l passing through p and in an elliptic geometry it is not possible to construct any lines through p parallel to l.

This paper gives an example of a geometry where more than one form of the parallel axiom is valid within the geometry.

A Smarandache geometry is a geometry in which at least one of the five fundamental axioms is either validated and invalidated, or only invalidated, but in multiple ways (in the same geometric space).

Consider the parallel or the fifth axiom of Euclidean geometry.

Through a point outside a line one can only draw one parallel to that line.

We can build a model of Smarandache geometry where the axiom of parallels is validated for some lines and points, and invalidated for other points and lines.

Figure 1

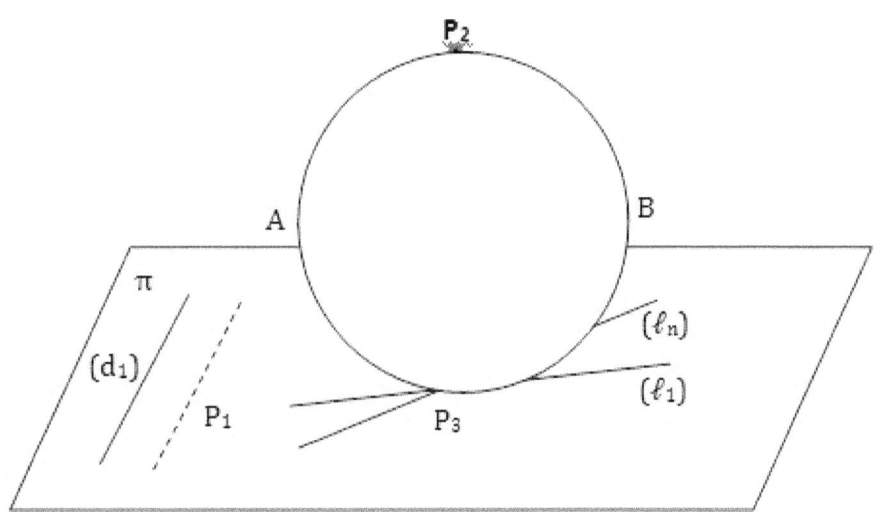

Consider a plane (π) and a sphere S of center O that is tangent to the plane (π) in the point P_3. The line (d_1) and the point P_1 belong to the plane (π).

The concepts of "line" and "point" in the plane (π) are the classical ones. On the sphere, the "line" is a big circle of the sphere, and the "point" is any classical point on the surface of the sphere.

Two lines are called parallel if they have no common point. Hence, the parallel axiom has three different forms in this Smarandache geometry model:

1. Through the point P_1 one can draw only one parallel to the line (d_1) [as in the Euclidean geometry].

2. Through the point P_2 one cannot draw any parallel to the line AB because a great circle of the sphere passing through P_2 will intersect the great circle AB [as in the non-Euclidean elliptical geometry].

3. Through the point P_3 belonging to the plane (π) and to the sphere S, one can draw an infinity of lines (ℓ_1), ..., (ℓ_n), ..., all contained in the plane (π), which do not intersect the line AB, so they are parallel to the line AB [as in the non-Euclidean hyperbolic geometry].

References

[1] Linfan Mao, *Automorphism groups of maps, surfaces and Smarandache geometries*, 2005, http://xxx.lanl.gov/pdf/math/0505318v1

[2] D. Rabounski, *Smarandache Spaces as a New Extension of the Basic Space-Time of General Relativity*, **Progress in Physics**, 2010, Vol: 2, Issue: Pages/record No.: L1-L2, DOAJ record, Sweden.

[3] Jerry L. Brown, *The Smarandache Counter-Projective Geometry*, Abstracts of Papers Presented to the American Mathematical Society Meetings, Vol 17, No. 3, Issue 105, 595, 1996.

[4] Sandy P. Chimienti, Mihaly Bencze, *Smarandache Anti-Geometry*, **Bulletin of Pure and Applied Sciences**, Delhi, India, Vol. 17E, No. 1, pp. 103-114, 1998.

SMARANDACHE'S CONCURRENT LINES THEOREM

Edited by Dr. M. Khoshnevisan

Neuro Intelligence Center, Australia

E-mail: mkhoshnevisan@neurointelligence-center.org

Abstract

In this paper we present the Smarandache's Concurrent Lines Theorem in the geometry of the polygon. The theorem states that if a polygon having any number of sides greater than 3 is circumscribed around a circle then the set of lines connecting vertices of the polygon in combination with the set of lines connecting points of tangency to the circle has at least three of the lines are concurrent.

Let's consider a polygon (which has at least four sides) circumscribed to a circle, and D the set of its diagonals and the lines joining the points of contact of two non-adjacent sides. Then the set D contains at least three concurrent lines.

Proof:

Let n be the number of sides. If $n = 4$, then the two diagonals and the two lines joining the points of contact of two adjacent sides are concurrent (according to Newton's Theorem).

The case $n > 4$ is reduced to the previous case: we consider any polygon $A_1...A_n$ (see the figure)

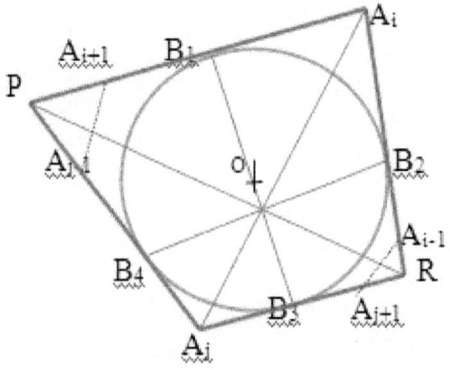

circumscribed to the circle and we choose two vertices A_i, A_j ($i \neq j$) such that

$$A_j \, A_{j-1} \cap A_i \, A_{i+1} = P$$

and

$$A_j \, A_{j+1} \cap A_i \, A_{i-1} = R.$$

32

Let B_h, $h \in \{1,2,3,4\}$, be the contact points of the quadrilateral PA_jRA_i with the circle of center O. Because of the Newton's theorem, the lines A_iA_j, B_1B_3 and B_2B_4 are concurrent.

Open Problems related to the Smarandache Concurrent Lines Theorem

1) In what conditions are there more than three concurrent lines?

2) What is the maximum number of concurrent lines that can exist (and in what conditions)?

3) What about an alternative of this problem: to consider instead of a circle an ellipse, and then a polygon *ellipsoscribed* (let's invent this word, *ellipso-scribed*, meaning a polygon whose all sides are tangent to an ellipse which inside of it): how many concurrent lines we can find among its diagonals and the lines connecting the point of contact of two non-adjacent sides?

4) What about generalizing this problem in a 3D-space: a sphere and a polyhedron circumscribed to it?

5) Or instead of a sphere to consider an ellipsoid and a polyhedron *ellipsoido-scribed* to it?

6) What about considering the lines that connect a vertex with a non-adjunct point of contact? Are there three or more such lines that intersect in the same point? (Consider all previous five questions.)

Comments

Of course, we can go by construction reversely: take a point inside a circle (similarly for an ellipse, a sphere, or ellipsoid), then draw secants passing through this point that intersect the circle (ellipse, sphere, ellipsoid) into two points, and then draw tangents to the circle (or ellipse), or tangent planes to the sphere or ellipsoid) and try to construct a polygon (or polyhedron) from the intersections of the tangent lines (or of tangent planes) if possible.

For example, a regular polygon (or polyhedron) has a higher chance to have more concurrent such lines.

In the 3D space, we may consider, as alternative to this problem, the intersection of planes (instead of lines).

References

[1] F. Smarandache, **Problèmes avec et sans… problèmes!,** Problem 5.36, p. 54, Somipress, Fés, Morocco, 1983.

[2] F. Smarandache, **Eight Solved and Eight Open Problems in Elementary Geometry**, in arXiv.org, Cornell University, NY, USA.

[3] M. Khoshnevisan, *Smarandache's Concurrent Lines*, NeuroIntelligence Center, Australia, http://www.scribd.com/doc/28325583/A-Smarandache-Concurrent-Lines-Theorem

THE CAREER SAVE PERCENTAGE PROFILE OF NHL GOALIES

Douglas A. Raeder

Paul M. Sommers

Department of Economics

Middlebury College

Middlebury, Vermont 05753

psommers@middlebury.edu

Abstract

Physical abilities of professional athletes eventually depreciate with age. The authors examine 75 goalies in the National Hockey League (NHL) who played at least five games in the 2008-09 season and how their save percentage varied from season to season throughout their career. The authors' regression model indicates that the save percentage profile of a representative NHL goalie reaches a peak around their sixth year in the NHL.

For most professional athletes, productivity increases rather quickly, tops off, and then declines as skills diminish or health declines with age. Career length and the point where productivity "tops off" varies between sports. And, even within a given sport, the point where one reaches his or her prime may not be well-defined.

In this brief note, we endeavor to show how a representative goalie's save percentage varies with years in the National Hockey League (NHL). To do this, we will regress the save percentage[1,2] of a goalie in the NHL against his career save percentage (through the 2008-09 season), years in the NHL, and years squared for all 75 goalies who played a minimum of five games in 2008-09. These seventy-five goalies had a combined total of 373 years of playing experience.[3] For a goalie (like Chris Osgood of the Detroit Red Wings) with, say, 15 years of playing experience (as of 2008-09), there would be 15 observations (of the 373 total) for that particular player. For a goalie with a given career save percentage, we should be able to show that the season-to-season save percentage rises at a diminishing rate for several seasons and reaches a peak. After this peak is reached, the save percentage declines. Of particular interest is the point where the performance profile peaks for an NHL goalie.

The Model

A goalie's save percentage in year t [SVP_t] for each of his n years in the NHL (where the player must play a minimum of ten games per season before 2008-09) was regressed against career year (which for a representative goalie varies from 1 to n) [$Year_t$], career years squared [$Year_t^2$], and his lifetime or career save percentage [*Lifetime_SVP*] through the 2008-09 season, as follows:

$$\text{SVP}_t = \beta_0 + \beta_1 \textbf{Year}_t + \beta_2 \textit{Year}_t^2 + \beta_3 \textbf{Lifetime\_SVP} + \varepsilon_t \qquad (1)$$

where ε_t denotes a stochastic disturbance (or error) term which may take on positive or negative

values. If a goalie's save percentage rises with career year and then falls after a point, then b_1,

the least squares estimate for β_1, should be positive and b_2, the least squares estimate for β_2,

should be negative.[4] The peak point is found by taking the partial derivative of SVP_t with respect

to $Year_t$, setting this derivative equal to zero, and solving for $Year_t$ in terms of b_1 and b_2. That is,

$$\frac{\partial SVP_t}{\partial Year_t} = b_1 + b_2\, Year_t = 0 \qquad (2)$$

or $Year^* = -\, b_1 / 2 b_2$, where $Year^*$ denotes the career year where save percentage

performance data are from www.nhl.com .

The Results

The 2008-09 regression equation is as follows (*t*-values in parentheses):

$$SVP_t = -\, 0.057 + .0009212\ Year_t - .000077\ Year_t^2 + 1.059\ Lifetime\_SVP \qquad (3)$$
$$ (-0.80) \qquad (1.83) \qquad\qquad (-2.15) \qquad\qquad (13.51)$$

$$R^2 = .345$$

For this regression, $Year^* = 5.98$, that is, the point in a goalie's career where the save percentage

peaks. Figure 1 shows the profile of a goalie whose career save percentage was equal to the

average of the seventy-five goalies in the sample (.904).

Concluding Remarks

The save percentage profile of a representative NHL goalie is derived for all goalies in 2008-09. The profile typically rises, at least up to a point, and then falls (sometime around their sixth year in the league).

Interested readers might be curious to see how this profile has changed (if at all) from one decade to the next. Are goalies peaking later in their career than did their counterparts one or two decades earlier? If NHL goalies now benefit more than their predecessors from off-season training and better conditioning, not to mention the improvements in sports medicine and physical therapy, are performance profiles now flatter or higher beyond the peak point than they used to be?

Figure 1.

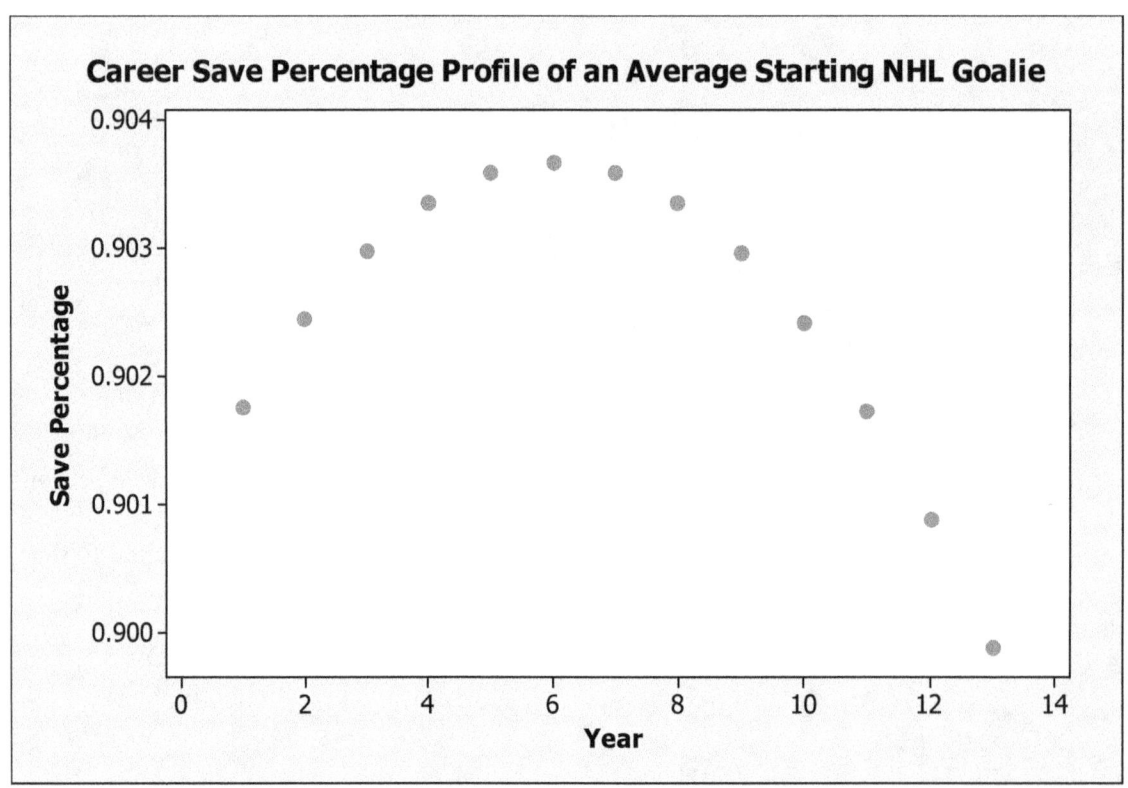

Footnotes

1. A goalie's save percentage is found by dividing the number of saves (i.e., the number of shots on goal a goaltender stops) by the total number of shots on goal. A 90 percent save percentage is here expressed as .900.

2. Goals against average (another common measure of a goalie's performance) is heavily influenced by his teammates' defensive efforts. Save percentage is considered to be a better measure of the goalie's own performance in net.

3. In seasons prior to 2008-09, we required the goalie to play a minimum of ten games during the season to count as a "year of playing experience."

4. The coefficient on *Lifetime_SVP* should be close to 1.0. The estimated coefficient b_3 could be greater than 1.0 since seasons with fewer than 10 games played (usually in a rookie's first year or two in the NHL with typically little ice time) can result in uncharacteristically good save percentages and hence lead to an artificially high goalie's lifetime save percentage.

SOME UNSOLVED PROBLEMS IN NUMBER THEORY

Taken from **Only Problems, Not Solutions!,** by Florentin Smarandache, Chicago, 1991, 1993.

1) A number is *pseudo-prime* if some permutation of its digits, including the identity permutation, is a prime. Of course all primes are pseudo-primes.

For example 14 is a pseudo-prime since a permutation of its digits, 41, is a prime.

Now let's consider the infinite sequence of primes and perform the same non-identity permutation of digits for each prime of two or more digits. Does that sequence contain an infinite number of primes?

2) A number is a *pseudo-square* if some permutation of its digits, including the identity permutation, is a square. Of course all squares are pseudo-squares.

For example 52 is a pseudo-square since a permutation of its digits, 25, is a square.

Now let's consider the infinite sequence of squares and perform the same non-identity permutation of digits for each square of two or more digits. Does that sequence contain an infinite number of squares?

3) Consider the following *binary sieve*:

Start with the set of natural numbers

1, 2, 3, 4, 5, . . .

a) Remove every second number from this list

b) Remove every fourth number from what remains

c) Remove every eighth number from what remains

. . . .

i) Remove every 2^k number from what remains

 Repeat to infinity

It is clear that there will be an infinite number of numbers remaining when this process is complete. The question becomes:

Are there infinitely many primes in this sequence?

You can download free e-books of number theory from the **Digital Library of Science**:

http://www.gallup.unm.edu/~smarandache/eBooks-otherformats.htm

ALTERNATING ITERATIONS OF THE SUM OF DIVISORS FUNCTION AND THE PSEUDO-SMARANDACHE FUNCTION

Henry Ibstedt

Glimminge 2036

28060 Broby

Sweden

henry.ibstedt@gmail.com

Abstract

This study is an extension of work done by Charles Ashbacher [3]. Iteration results have been re-defined in terms of invariants and loops. Further empirical studies and analysis of results have helped throw light on a few intriguing questions.

Introduction

The following definition forms the basis of Ashbacher's study:

For n>1, the Zσ sequence is the alternating iteration of the Sum of Divisors Function σ followed by the Pseudo-Smarandache function Z.

The Zσ sequence originated by n creates a cycle. Ashbacher identified four 2 cycles and one 12 cycle. These are listed in table 1.

Table 1
Iteration cycles C_1 - C_5.

n	C_k	Cycle
2	C_1	$3 \leftrightarrow 2$
$3 \leq n \leq 15$	C_2	$24 \leftrightarrow 15$
n=16	C_3	$31 \rightarrow 32 \rightarrow 63 \rightarrow 104 \rightarrow 64 \rightarrow 127 \rightarrow 126 \rightarrow 312 \rightarrow 143 \rightarrow 168 \rightarrow 48 \rightarrow 124$
$17 \leq n \leq 19$	C_2	$24 \leftrightarrow 15$
n=20	C_3	$42 \leftrightarrow 20$
n=21	C_3	$31 \rightarrow 32 \rightarrow 63 \rightarrow 104 \rightarrow 64 \rightarrow 127 \rightarrow 126 \rightarrow 312 \rightarrow 143 \rightarrow 168 \rightarrow 48 \rightarrow 124$
$22 \leq n \leq 24$	C_2	$24 \leftrightarrow 15$
n=25	C_3	$31 \rightarrow 32 \rightarrow 63 \rightarrow 104 \rightarrow 64 \rightarrow 127 \rightarrow 126 \rightarrow 312 \rightarrow 143 \rightarrow 168 \rightarrow 48 \rightarrow 124$
n=26	C_3	$42 \leftrightarrow 20$
...		
n=381	C_5	$1023 \leftrightarrow 1536$

The search for new cycles was continued up to n=552,000. No new ones were found. This led Ashbacher to pose the following questions

1) Is there another cycle generated by the $Z\sigma$ sequence?

2) Is there an infinite number of numbers n that generate the two cycle $42 \leftrightarrow 20$?

3) Are there any other numbers n that generate the two cycle $2 \leftrightarrow 3$?

4) Is there a pattern to the first appearance of a new cycle?

Ashbacher concludes his article by stating that these problems have only been touched upon and encourages others to further explore these problems.

An extended study of the $Z\sigma$ iteration

It is amazing that hundreds of thousands of integers subjected to a fairly simple iteration process all end up with final results that can be described by a few small integers. This merits a closer analysis. In an earlier study of iterations [2] the iteration results were classified in terms of invariants, loops and divergents. Applying the iteration to a member of a loop produces another member of the same loop. The cycles described in the previous section are not loops. The members of a cycle are not generated by the same process, half of them are generated by $Z(\sigma(Z(\ldots\sigma(n)\ldots)))$ while the other half is generated by $(\sigma(Z(\ldots\sigma(n)\ldots)))$, i.e. we are considering two different operators. This leads to a situation were the iteration process applied to a member of a cycle may generate a member of another cycle as described in table 2.

This situation makes it impossible to establish a one-to-one correspondence between a number n to which the sequence of iterations is applied and the cycle that it will generate. Henceforth the iteration function will be $Z(\sigma(n))$ which will be denoted $Z\sigma(n)$ while results included in the above cycles originating from $\sigma(Z(\ldots\sigma(n)\ldots))$ will be considered as intermediate elements. This leads to an unambiguous situation which is shown in table 3.

Table 2

A $Z\sigma$ iteration applied to an element belonging to one cycle may generate an element belonging to another cycle.

	C_1		C_2		C_3		C_4												C_5	
n	2	3	15	24	20	42	31	32	63	104	64	127	126	312	143	168	48	124	1023	1536
$\sigma(n)$	3	4	24	60	42	96	32	63	104	210	127	128	312	840	168	480	124	224	1536	4092
$Z(\sigma(n))$	2	7	15	15	20	63	63	27	64	20	126	255	143	224	48	255	31	63	1023	495
$\sigma(Z(\sigma(n)))$		8						40				...		504		...				936
$Z(\sigma(Z(\sigma(n))))$		15						15				15		63		15				143
...																				
Generates	C_1	C_2	C_2	C_2	C_3	C_4	C_4	C_2	C_4		C_3	C_4	C_2	C_4	C_4	C_4	C_2	C_4	C_5	C_4
*=Shift to other cycle		*				*		*				*		*		*				*

Table 3

The Zσ iteration process described in terms of invariants, loops and intermediate elements.

	I_1	I_2	I_3			Loop				I_4
n	2	15	20	31	63	64	126	143	48	1023
$Z(\sigma(n))$	2	15	20	63	64	126	143	48	31	1023
Intermediate element	3	24	42	32	104	127	312	168	124	1536

We have four invariants I_1, I_2, I_3 and I_4 and one loop L with six elements. No other invariants or loops exist for $n \le 10^6$. Each number $n \le 10^6$ corresponds to one of the invariants or the loop. The distribution of results of the Zσ iteration has been examined by intervals of size 50000 as shown in table 4. The stability of this distribution is amazing. It deserves a closer look and will help bringing us closer to answers to the four questions posed by Ashbacher.

Question 3: Are there any other numbers n that generate the two cycle $2 \leftrightarrow 3$? In the framework set for this study this question will reformulated to: Are there any numbers other than $n = 2$ that belongs to the invariant 2?

Theorem: $n = 2$ is the only element for which $Z(\sigma(n)) = 2$.

Proof:

$Z(x) = 2$ has only one solution which is x=3. $Z(\sigma(n)) = 2$ can therefore only occur when $\sigma(n)=3$ which has the unique solution n=2.

Question 1: Is there another cycle generated by the $Z\sigma$ sequence?

Question 2: Are there an infinite number of numbers n that generate the two cycle $42 \leftrightarrow 20$?

Conjecture: There are infinitely many numbers n which generate the invariant 20 (I_3).

Support: Although the statistics shown in table 4 only skims the surface of the "ocean of numbers" the number of numbers generating this invariant is as stable as for the other invariants and the loop. To this is added the fact that any number $n > 10^6$ will either generate a new invariant or loop (highly unlikely) or "catch on to" one of the already existing end results where I_4 will get its share as the iteration "filters through" from 10^6 until it gets locked onto one of the established invariants or the loop.

Discussion: The search up to $n = 10^6$ revealed no new invariants or loops. If another invariant or loop exists it must be initiated by a number $n > 10^6$. Let N be the value of n up to which the search has been completed. For n=N+1 there are three possibilities:

Possibility 1: $Z(\sigma(n)) \leq N$. In this case continued iteration repeats iterations which have already been done in the complete search up to $n = N$. No new loops or invariants will be found.

Possibility 2: $Z(\sigma(n)) = n$. If this happens then $n = N+1$ is a new invariant. A necessary condition for an invariant is therefore that

$$\frac{n(n+1)}{2\sigma(n)} = q, \text{ where q is a positive integer} \tag{1}$$

. . .

If, in addition **no** m < n exists such tha

Table 4

Zσ iteration iteration results.

Interval	I_2	I_3	Loop	I_4
3-50000	18824	236	29757	1181
50001-100000	18255	57	30219	1469
100001-150000	17985	49	30307	1659
150001-200000	18129	27	30090	1754
200001-150000	18109	38	30102	1751
250001-300000	18319	33	29730	1918
300001-350000	18207	24	29834	1935
350001-400000	18378	18	29622	1982
400001-450000	18279	21	29645	2055
450001-550000	18182	24	29716	2078
500001-550000	18593	18	29227	2162
550001-600000	18159	19	29651	2171
600001-650000	18596	25	29216	2163
650001-700000	18424	26	29396	2154
700001-750000	18401	20	29409	2170
750001-800000	18391	31	29423	2155
800001-850000	18348	22	29419	2211
850001-900000	18326	15	29338	2321
900001-950000	18271	24	29444	2261
950001-1000000	18517	31	29257	2195
Average	18335	38	29640	1987

There are 111 potential invariant candidates for n up to $3 * 10^8$ that satisfy the necessary condition given in (1).   Only four of them n = 2, 15, 20 and 1023 satisfied condition (2). It seems that for a given solution to (1) there is always, for $n > N > 1023$, a solution to (2) with $m < n$. This is plausible since we know [4] that $\sigma(n) = O(n^{1+\delta})$ for every positive δ which means that $\sigma(n)$ is small compared to $n(n+1) \approx n^2$ for large n.

Example: The largest $n < 3 * 10^8$ for which (1) is satisfied is n=292,409,999 with $\sigma(292,409,999) = 341145000$ and $292409999 * 292410000/(2 * 341145000) = 125318571$. But m = 61370000 < n exists for which $61370000 * 61370001/(2 * 341145000) = 5520053$, an integer, which means that n is not invariant.

<u>**Possibility 3:**</u> $Z(\sigma(n)) > N$. This could lead to a new loop or invariant. Let's suppose that a new loop of length $k \geq 2$ is created. All elements of this loop must be greater than N otherwise the iteration sequence would fall below N and end up on a previously known invariant or loop. A necessary condition for a loop is therefore that

$$Z(\sigma(n)) > n \text{ and } Z(\sigma(Z(\sigma(n)))) \geq n. \tag{3}$$

Denoting the kth iteration $(Z\sigma)_k(n)$ we must finally have

$$(Z\sigma)_k(n) = (Z\sigma)_j(n) \text{ for some } k \neq j, \text{ interpreting } (Z\sigma)_0(n) = n. \tag{4}$$

There isn't much hope for all this to happen since, for large n, already $Z(\sigma(n)) > n$ is a scarce event and becomes scarcer as we increase n. A study of the number of incidents where

49

$(Z\sigma)_3(n) > n$ for n < 800,000 was made. There are only 86 of them, of these 65 occurred for

n < 100,000. From n = 510,322 to n = 800,000 there was not a single incident.

Question 4: No particular patterns were found.

Epilog

In empirical studies of numbers the search for patterns and general behaviors is an interesting

and important part. In this iteration study it is amazing that all these numbers, where not even the

sky is the limit[1], after a few iterations filter down to end up on one of three invariants or a single

loop. The other amazing thing is the relative stability of distribution between the three invariants

and the loop with increasing n (see table 4). When $(Z\sigma)_k(n)$ drops below n it catches on to an

integer which has already been iterated and which has therefore already been classified to belong

to one of the four terminal events. This in my mind explains the relative stability. In general the

end result is obtained after only a few iterations. It is interesting to see that $\sigma(n)$ often assumes

the same value for values of n which are fairly close together. Here is an example: $\sigma(n)=3024$ for

n=1020, 1056, 1120, 1230, 1284, 1326, 1420, 1430, 1484, 1504, 1506, 1564, 1670, 1724, 1826,

1846, 1886, 2067, 2091, 2255, 2431, 2515, 2761, 2839, 2911, 3023. I may not have brought this

subject much further but I hope to have contributed some light reading in the area of recreational

mathematics.

---

[1] "Not even the sky is the limit" expresses the same dilemma as the title of the authors book "Surfing on the ocean of numbers". Even with for ever faster computers and better software for handling large numbers empirical studies remain very limited.

References

1. H. Ibstedt, **Surfing on the Ocean of Numbers**, Erhus University Press, 1997.

2. Charles Ashbacher, **Pluckings From the Tree of Smarandache Sequences and Functions**, American Research Press, 1998.

3. Charles Ashbacher, "On Iterations That Alternate the Pseudo-Smarandache and Classic Functions of Number Theory", *Smarandache Notions Journal, Vol. 11, No 1-2-3.*

4. G.H. Hardy and E.M. Wright, **An Introduction to the Theory of Numbers**. Oxford University Press, 1938.

ALTERNATING ITERATIONS OF THE EULER ϕ FUNCTION AND THE PSEUDO-SMARANDACHE FUNCTION

Henry Ibstedt

Glimminge 2036

28060 Broby

Sweden

henry.ibstedt@gmail.com

Abstract

This study originates from questions posed on alternating iterations involving the Pseudo-Smarandache function $Z(n)$ and the Euler function $\phi(n)$. An important part of the study is a formal proof of the fact that $Z(n) < n$ for all $n \neq 2^k$ ($k \geq 0$). Interesting questions have been resolved through the surprising involvement of Fermat numbers.

The Behaviour of the Pseudo-Smarandache Function

Definition: The Pseudo-Smarandache function Z(n) is the smallest positive integer m such that 1+2+...+m is divisible by n.

Adding up the arithmetical series results in an alternative and more useful formulation: For a given integer n, Z(n) equals the smallest positive integer m such that $(m(m+1)) / 2n$ is an integer. Some properties and values of this function are given in [1], which also contains an effective computer algorithm for calculation of Z(n). The following properties are evident from the definition:

1. $Z(1) = 1$

2. $Z(2) = 3$

3. For any odd prime p, $Z(p^k) = p^k - 1$ for $k \geq 1$

4. For $n = 2^k$, $k \geq 1$, $Z(2^k) = 2^{k+1} - 1$

We note that $Z(n) = n$ for $n = 1$ and that $Z(n) > n$ for $n = 2^k$ when $k \geq 1$. Are there other values of n for which $Z(n) \geq n$? The answer is no, there are none, but to my knowledge no proof has been given. Before presenting the proof it might be useful to see some elementary results and calculations on Z(n). Explicit calculations of $Z(3 *2^k)$ and $Z(5 * 2^k)$ have been carried out by Charles Ashbacher [2]. For $k > 0$:

$$
Z(5 * 2^k) = \begin{cases} 2^{k+2} & \text{if } k \equiv 0 \pmod 4 \\[2mm] 2^{k+1} & \text{if } k \equiv 1 \pmod 4 \\[2mm] 2^{k+2} - 1 & \text{if } k \equiv 2 \pmod 4 \\[2mm] 2^{k+1} - 1 & \text{if } k \equiv 3 \pmod 4. \end{cases}
$$

A specific remark was made in each case that $Z(n) < n$. In this study we will prove that $Z(n) < n$ for all $n \neq 2^k$, $k \geq 0$, but before doing so we will continue to study $Z(a * 2^k)$, for a odd and $k > 0$. In particular we will carry out a specific calculation for $n = 7 * 2^k$.

We look for the smallest integer m for which

$$
\frac{m(m + 1)}{7 * 2^{k+1}}
$$

is an integer. We distinguish two cases in table 1.

By choosing the smallest m in each case we find:

$$Z(7*2^k) = \begin{cases} 2^{k+1} - 1 & \text{if } k \equiv -1 \pmod{3} \\ \\ 3 * 2^{k+1} & \text{if } k \equiv 0 \pmod{3} \\ \\ 2^{k+2} - 1 & \text{if } k \equiv 1 \pmod{3} \end{cases}$$

again we note that $Z(n) < n$.

Table 1

Case 1:	Case 2:

$$m=7x$$

$$m=2^{k+1}y$$

$$m+1=2^{k+1}y$$

$$m+1=7x$$

Eliminating m results in

$$2^{k+1}y-1=7x$$

$$2^{k+1}y+1=7x$$

$$2^{k+1}y\equiv1 \pmod 7$$

$$2^{k+1}y\equiv-1 \pmod 7$$

Since $2^3\equiv1 \pmod 7$ we have

If $k\equiv-1 \pmod 3$ then

$$y\equiv1 \pmod 7 \ ; \ m=2^{k+1}-1$$ $$y\equiv8 \pmod 7; \ m=2^{k+1}\cdot8=2^{k+4}$$

If $k\equiv0 \pmod 3$ then

$$2y\equiv1 \pmod 7, \ y=4; \ m=2^{k+1}\cdot4-1=2^{k+3}-1$$ $$y\equiv3 \pmod 7; \ m=3\cdot2^{k+1}$$

If $k\equiv1 \pmod 3$ then

$$4y\equiv1 \pmod 7, \ y=2; \ m=2^{k+1}\cdot2-1=2^{k+2}-1$$ $$y\equiv5 \pmod 7; \ m=5\cdot2^{k+1}$$

In a study of alternating iterations [3] it is stated that apart from when $n = 2^k$ ($k \geq 0$) $Z(n)$ is <u>at most n</u>. If it ever happened that $Z(n) = n$ for $n > 1$ then the iterations of $Z(n)$ would arrive at an invariant, i.e. $Z(\ldots Z(n)\ldots) = n$. This cannot happen, therefore it is important to prove the following theorem.

Theorem: $Z(n) < n$ for all $n \neq 2^k$, $k \geq 0$.

Proof:

k > 0. Write n in the form $n = a * 2^k$, where $a \geq 3$ is odd and $k > 0$. Consider the following two cases:

1. $a \mid m$ and $2^{k+1} \mid (m + 1)$

2. $2^{k+1} \mid m$ and $a \mid (m + 1)$.

In case 1 we write $m = ax$. We then require $2^{k+1} \mid (ax + 1)$, which means that we are looking for solutions to the congruence

$$ax \equiv -1 \pmod{2^{k+1}} \tag{1}$$

In case 2 we write $m + 1 = ax$ and require $2^{k+1} \mid (ax - 1)$. This corresponds to the congruence

$$ax \equiv 1 \pmod{2^{k+1}} \tag{2}$$

These congruencies have solutions in the interval $0 \leq x < 2^{k+1}$ which contains a complete set of residues modulo 2^{k+1}. If $x = x_1$ is a solution to $ax + 1 \equiv 0 \pmod{2^{k+1}}$ then $x = x_2$ is a solution to $ax - 1 \equiv 0 \pmod{2^{k+1}}$ because we have

$$ax_2 - 1 = a(2^{k+1} - x_1) - 1 = a2^{k+1} - (ax_1 + 1) \equiv 0 \pmod{2^{k+1}}$$

Furthermore if one of the solutions is greater than 2^k then the other is smaller than 2^k because if $x_1 > 2^k$ then $x_2 = 2^{k+1} - x_1 < 2^{k+1} - 2^k = 2^k (2 - 1) = 2^k$. So we have $m = ax$ or $m = ax - 1$ with

$0 < x < 2^k$, which means that $m = \min(ax_1, ax_2 - 1) < n$ exists so that $(m(m+1))/2$ is divisible by n when $a \geq 3$ in $n = a * 2^k$.

k=0. In

$$\frac{m(m+1)}{2n}$$

we chose $m + 1 = n$ and consequently $m = n - 1$ and $m < n$.

Comments

Let´s call the m that we have proved to exist m_{max}. We then have $Z(n) \leq m_{max} < n$. The solution may be of another form than those used to prove the theorem, for example if $a = a_1a_2$ we could have $a_1 | m$ and $a_2 \, 2^{k+1} | (m + 1)$. Another possibility would be a $* \, 2^{k+1} | m$ or

$a * 2^{k+1} | (m + 1)$. When $k = 0$ we may have $n = a = a_1a_2$ with $a_1 | m$ and $a_2 | (m + 1)$. In fact values of $Z(n)$ will be found using the algorithm

$$Z(n) = \frac{-1 + \sqrt{1 + 8kn}}{2} \tag{3}$$

where k is the smallest positive integer which makes $Z(n)$ integer.

A few numerical examples are given in the form $(Z(n), m_{max}, n)$ for n in two arbitrarily chosen intervals $100 \leq n \leq 110$ and $190 \leq n \leq 200$, primes excluded.

(24,25,100), (51,51,102),(64,65,104),(14,104,105),(52,53,106),(80,81,108),(44,55,110).

(19,95,190),(128,129,192),(96,97,194),(39,194,195),(48,49,196),(44,99,196),(175,175,200).

Here is another theorem which as a corollary proves that $Z(n)$ is not a multiplicative function. The proof[2] is left to the reader.

Theorem: If $n = pq$, where p and q are two distinct primes with $g = q - p$, then

$$Z(n) = \min\{p(qk + 1) / g \text{ where } pk + 1 \equiv 0 \pmod{g}, q(pk - 1) / g \text{ where } pk - 1 \equiv 0 \pmod{g}\}.$$

Iterating the Pseudo-Smarandache Function

The theorem proved in the previous section assures that an iteration of the pseudo-Smarandache function does not result in an invariant, i.e. $Z(n) \neq n$ is true for $n \neq 1$. On iteration the function will leap to a higher value only when $n = 2^k$. It can only go into a loop (or cycle) if after one or more iterations it returns to 2^k. Up to $n = 2^{28}$ this does not happen and a statistical view on the results displayed in table 2 makes it reasonable to conjecture that it never happens. Each row in table 2 corresponds to a sequence of iterations starting on $n = 2^k$ finishing on the final value 2. The largest number of iterations required for this was 24 and occurred for $n = 2^{14}$ which also had the largest numbers of leaps form 2^j to $2^{j+1} - 1$. Leaps are represented by ↑ in table 2. For $n = 2^{11}$ and 2^{12} the iterations are monotonic decreasing.

Iterating the Euler ϕ Function

The function $\phi(n)$ is defined for $n>1$ as the number of positive integers less than and prime to n. The analytical expression is given by

---

[2] For the proof see *Surfing on the Ocean of Numbers.* H. Ibstedt (1997)

$$\phi(n) = n \prod_{p|n} \left(1 - \frac{1}{p}\right)$$

For n expressed in the form $n = p_1^{\alpha_1} p_2^{\alpha_2} \cdots p_r^{\alpha_r}$ it is often useful to express $\phi(n)$ in the form

$$\phi(n) = p_1^{\alpha_1 - 1}(p_1 - 1) p_2^{\alpha_2 - 1}(p_2 - 1) \cdots p_r^{\alpha_r - 1}(p_r - 1)$$

It is obvious from the definition that $\phi(n) < n$ for all $n > 1$. Applying the ϕ function to $\phi(n)$ we

will have $\phi(\phi(n)) < \phi(n)$. After a number of such iterations the end result will of course be 1. It is

what this chain of iterations looks like which is interesting and will be studied here. For

convenience we will write $\phi_2(n)$ for $\phi(\phi(n))$. $\phi_k(n)$ stands for the kth iteration. To begin with we

will look at the iteration of a few prime powers

$\phi(2^\alpha) = 2^{\alpha-1}$, $\phi_k(2^\alpha) = 2^{\alpha-k}$, ... $\phi_\alpha(2^\alpha) = 1$.

$\phi(3^\alpha) = 3^{\alpha-1} * 2$, $\phi_2(3^\alpha) = 3^{\alpha-2} * 2$, ... $\phi_k(3^\alpha) = 3^{\alpha-k} * 2$ for $k \leq \alpha$.

In particular $\phi_\alpha(3^\alpha) = 2$.

Proceeding in the same way we will write down $\phi_k(p^\alpha)$, $\phi_\alpha(p^\alpha)$ and the first occurrence of an

iteration result which consists purely of a power of 2.

$\phi_k(5^\alpha) = 5^{\alpha-k} * 2^{k+1}, k \leq \alpha$ $\qquad$ $\phi_\alpha(5^\alpha) = 2^{\alpha+1}.$

$\phi_k(7^\alpha) = 7^{\alpha-k} * 3 * 2^k, k \leq \alpha$ $\qquad$ $\phi_\alpha(7^\alpha) = 3 * 2^\alpha$ $\qquad$ $\phi_{\alpha+1}(7^\alpha) = 2^\alpha.$

$\phi_k(11^\alpha) = 11^{\alpha-k} * 5 * 2^{2k-1}, k \leq \alpha$ $\qquad$ $\phi_\alpha(11^\alpha) = 5 * 2^{2\alpha-1}$ $\qquad$ $\phi_{\alpha+1}(11^\alpha) = 2^{2\alpha}.$

$\phi_k(13^\alpha) = 13^{\alpha-k} * 3 * 2^{2k}, k \leq \alpha$ $\qquad$ $\phi_\alpha(13^\alpha) = 3 * 2^{2\alpha}$ $\qquad$ $\phi_{\alpha+1}(13^\alpha) = 2^{2\alpha}.$

$\phi_k(17^\alpha) = 17^{\alpha-k} * 2^{3k+1}, k \leq \alpha$ $\qquad$ $\phi_\alpha(17^\alpha) = 2^{3\alpha+1}.$

$\phi_k(19^\alpha) = 19^{\alpha-k} * 3^{k+1} * 2^k, k \leq \alpha$ $\qquad$ $\phi_\alpha(19^\alpha) = 3^{\alpha+1} * 2^\alpha$ $\qquad$ $\phi_{2\alpha+1}(19^\alpha) = 2^\alpha.$

$\phi_k(23^\alpha) = 23^{\alpha-k} * 11 * 5 * 2^{3k-4}, k \leq \alpha$ $\quad$ $\phi_\alpha(23^\alpha) = 11 * 5 * 2^{3\alpha-4}$ $\qquad$ $\phi_{\alpha+2}(23^\alpha) = 2^{3\alpha-1}.$

The characteristic tail of descending powers of 2 applies also to the iterations of composite integers and plays an important role in the alternating Z - ϕ iterations which will be subject of the next section.

Table 2

k/j	28	27	26	25	24	23	22	21	20	19	18	17	16	15	14	13	12	11	10	9	8	7	6	5	4	3	2
2																											↑
3																										↑	↑
4																									↑		↑
5																								↑		↑	↑
6																							↑			↑	↑
7																						↑					↑
8																					↑					↑	↑
9																				↑							↑
10																			↑					↑		↑	↑
11																		↑									
12																	↑										
13																↑										↑	↑
14															↑				↑					↑		↑	↑
15														↑													↑
16													↑														↑
17												↑								↑							↑
18											↑																↑
19										↑																↑	↑
20									↑																	↑	↑
21								↑																		↑	↑
22							↑																				↑
23						↑																				↑	↑
24					↑																						↑
25				↑																							↑
26			↑																						↑		↑
27		↑																								↑	↑
28	↑																							↑		↑	↑

The Alternating Iteration of the Euler ϕ Function Followed by the Smarandache Z Function.

Charles Ashbacher [3] found that the alternating iteration $Z(\ldots(\phi(Z(\phi(n))))\ldots)$ ends in

2-cycles of which he found the following four[3]:

2-cycle	First Instance
2 - 3	$3 = 2^2 - 1$
8 - 15	$15 = 2^4 - 1$
128 - 255	$255 = 2^8 - 1$
32768 - 65535	$65535 = 2^{16} - 1$

The following questions were then posed:

1) Does the Z - ϕ sequence always reduce to a 2-cycle of the form

$$2^{2^r - 1} \leftrightarrow 2^{2^r} - 1$$

for $r \geq 1$?

---

[3] It should be noted that 2, 8, 128 and 32768 can be obtained as iteration results only through iterations of the type $\phi(\ldots(Z(\phi(n)))\ldots)$ whereas the "complete" iterations $Z(\ldots(\phi(Z(\phi(n)))\ldots)$ lead to the invariants 3, 15, 255, 65535. Consequently we note that for example $Z(\phi(8))=7$ not 15, i.e. 8 does not belong to its own cycle.

2) Do any additional patterns always appear first for?

$$n = 2^{2^r} - 1$$

Theorem: The alternating iteration $Z(\ldots(\phi(Z(\phi(n))))\ldots)$ ultimately leads to one of the following five 2-cycles: 2 - 3, 8 - 15, 128 - 255, 32768 - 65535, 2147483648 - 4294967295.

Proof:

Since $\phi(n) < n$ for all $n > 1$ and $Z(n) < n$ for all $n \neq 2^k$ ($k \geq 0$) any cycle must have a number of the form 2^k at the lower end and $Z(2^k) = 2^{k+1} - 1$ at the upper end of the cycle. In order to have a 2-cycle we must find a solution to the equation

$$\phi(2^{k+1} - 1) = 2^k$$

If $2^{k+1} - 1$ were a prime $\phi(2^{k+1} - 1)$ would be $2^{k+1} - 2$ which solves the equation only when $k = 1$. A necessary condition is therefore that $2^{k+1} - 1$ is composite,

$2^{k+1} - 1 = f_1 * f_2 * \ldots * f_i * \ldots * f_r$, and that the factors are such that $\phi(f_i) = 2^{u_i}$ for $1 \leq i \leq r$. This means that each factor f_i must be a prime number of the form $2^{u_i} + 1$ and leads us to consider

$$q(r) = (2 - 1)(2 + 1)(2^2 + 1)(2^4 + 1)(2^8 + 1) \ldots (2^{2^{r-1}} + 1)$$

or

$$q(r) = (2^{2^r} - 1)$$

64

Numbers of the form $F_r = 2^{2^r} + 1$ are known as Fermat numbers. The first five of these are prime numbers

$F_0 = 3$, $F_1 = 5$, $F_2 = 17$, $F_3 = 257$, $F_4 = 65537$

while $F_5 = 641 * 6700417$ and F_6, F_7, F_8, F_9, F_{10} and F_{11} are all known to be composite. From this we see in equation 3

$$\phi(2^{2^r} - 1) = \phi(q(r)) = \phi(F_0)\phi(F_1) * \ldots * \phi(F_{r-1}) = 2 * 2^2 \cdots * 2^{2^{r-1}} = \tag{3}$$

$$2^{1+2+2^2+2^3+\ldots+2^{r-1}} = 2^{2^r - 1}$$

for r = 1, 2, 3, 4, 5 but breaks down for r = 6 (because F_5 is composite) and consequently also for r > 6.

Table 3 contains iterations of p^6. A horizontal line marks where the rest of the iterated values consist of descending powers of 2. (The obvious power of 2 tails have been omitted.)

Table 3

#	p=2	p=3	p=5	p=7	p=11	p=13	p=17	p=19	p=23
1	32	486	12500	100842	1610510	4455516	22717712	44569782	141599546
2	16	162	5000	28812	585640	1370928	10690688	14074668	61565020
3	8	54	2000	8232	212960	421824	5030912	4444632	21413920
4	4	18	800	2352	77440	129792	2367488	1403568	7448320
5	2	6	320	672	28160	39936	1114112	443232	2590720
6		2	128	192	10240	12288	524288	139968	901120
7			64	64	4096	4096	262144	46656	327680
8			32	32	2048	2048	131072	15552	131072
9			16	16	1024	1024	65536	5184	65536
10			8	8	512	512	32768	1728	32768
11			4	4	256	256	16384	576	16384
12			2	2	128	128	8192	192	8192
13					64	64	4096	64	4096

Evaluating equation (3) for r = 1, 2, 3, 4, 5 gives the complete table of expressions for the five 2-cycles that appear in table 4.

Table 4

Cycle #	2-cycle	Equiv.expression
1	$2 \leftrightarrow 3$	$2 \leftrightarrow 2^2 - 1$
2	$8 \leftrightarrow 15$	$2^3 \leftrightarrow 2^4 - 1$
3	$128 \leftrightarrow 255$	$2^7 \leftrightarrow 2^8 - 1$
4	$32768 \leftrightarrow 65535$	$2^{15} \leftrightarrow 2^{16} - 1$
5	$2147483648 \leftrightarrow 4294967295$	$2^{31} \leftrightarrow 2^{32} - 1$

The answers to the two questions are implicit in the previous theorem.

1) The Z-ϕ sequence always reduces to a 2-cycle of the form

$$2^{2^r - 1} \leftrightarrow 2^{2^r} - 1$$

for $r \geq 1$.

2) Only five patterns exist and they always appear first for $n = 2^{2^r} - 1$, r = 1, 2, 3, 4, 5.

A statistical survey of the frequency of the different 2-cycles, displayed in table 4, indicates that the lower cycles are favored when the initiating numbers grow larger. Cycle #4 could have appeared in the third interval but as can be seen it is generally scarcely represented. Prohibitive computer execution times made it impossible to systematically examine an interval were cycle #5 members can be assumed to exist. However, apart from the "founding member" 2147483648 ↔ 4294967295 a few individual members were calculated by solving the equation:

$$Z(\phi(n))=2^{32} - 1$$

The result is shown in table 5.

Table 5

The Distribution of Cycles for A Few Intervals of Length 1000

Interval	Cycle #1	Cycle #2	Cycle #3	Cycle #4
$3 \leq n \leq 1002$	572	358	70	-
$10001 \leq n \leq 11000$	651	159	190	-
$100001 \leq n \leq 101000$	759	100	141	0
$1000001 \leq n \leq 1001000$	822	75	86	17
$10000001 \leq n \leq 100001000$	831	42	64	63
$100000001 \leq n \leq 1000001000$	812	52	43	93

References

1. H. Ibstedt, **Surfing on the Ocean of Numbers**, Erhus University Press, 1997.

2. Charles Ashbacher, **Pluckings From the Tree of Smarandache Sequences and Functions**, American Research Press, 1998.

3. Charles Ashbacher, "On Iterations That Alternate the Pseudo-Smarandache and Classic Functions of Number Theory", *Smarandache Notions Journal, Vol. 11, No 1-2-3.*

BOOK REVIEWS

Edited by Charles Ashbacher

Charles Ashbacher Technologies

5530 Kacena Ave

Marion, IA 52302

E-mail: cashbacher@yahoo.com

In honor of Joseph Madachy and Martin Gardner, the first two reviews are of items that they authored.

Madachy's Mathematical Recreations, by Joseph S. Madachy, Dover Publications, Mineola, NY , 1979. 256 pp., $23.00 (paper). ISBN 978-0486237626

Recreational mathematics may be a misnomer to some, but to practitioners, it is an addiction. Whether you are creating new puzzles or solving those made by others, the thrill of the chase and the pride in solution make it an interesting and exciting hobby. A precise definition of recreational mathematics is impossible, often being a matter of personal taste. However, there are some areas that definitely satisfy the criteria, and in this book, Joseph Madachy, the longtime editor of Journal of Recreational Mathematics describes them and gives example problems. Many of these topics have been around for centuries without losing any of their charm.
The logic puzzle is very old, with the riddle of the sphinx being a part of Greek mythology. Origami, the Japanese art of paper folding, has been around for centuries. Although the folding of paper to make new designs is covered in this book, origami as it is normally defined is not. The most famous commentator on magic squares is none other than Benjamin Franklin, the original American polymath. The properties of numbers have been used as a point of mysticism and religion from the time of the ancient Greeks. Although Madachy clearly has a great deal of respect for numbers, he never descends into the farce of numerology. Alphametics, or the expression of an arithmetic expression as a collection of words, is a popular form of problem and can be used as a source of problems for courses in beginning algebra.
Problems and puzzles, whether they are crossword, word search, character shifting or encryption codes, are very popular, appearing regularly in most newspapers and magazines. This verifies the enormous human capacity and interest in exercising our brains. In this book, there are many problems, all solved, which will generate an increase in the activity and efficiency of your cerebral neurons.

Charles Ashbacher

Knots, Borromean Rings, Rep-tiles and Eight Queens, by Martin Gardner, Cambridge University Press, New York, New York, 2014. 274 pp., $17.99(paper). ISBN 978-0521758710

 When describing the works and career of Martin Gardner it is easy to run out of superlatives. At that point you either have to invent new ones or engage in recycling. The latter is what all people interested in mathematics should do, read and re-read the writings of Gardner. No one was more skilled in making mathematics understandable and he was the most influential person in lighting the fire of mathematical interest in people.

 This book is a collection of twenty articles that appeared in "Scientific American" in the years 1961, 1962 and 1963. All of them have the distinctive Gardner style of dissecting (sometimes quite literally in the case of geometry) a problem down into the fundamental components and then building the solution.

 One of the hardest jobs that can be envisioned would be to select a "best of" collection of Gardner's works. The following line that I used in a previous review has been quoted in other media outlets, "If there were mathematics of watching paint dry, Martin Gardner could make it interesting." As is always the case, this book is an opportunity to read and learn from the master.

Charles Ashbacher

Mathematicians Fleeing from Nazi Germany, by Reinhard Siegmund-Schultze, Princeton NJ, Princeton University Press, 2009, 504 pp. $49.50 (paper), ISBN 978-0-691-14041-4

 Not exactly light reading, this work is another excellent source for those interested in the history of mathematics in the twentieth century, a worthy companion to Sanford Segal's *Mathematician under the Nazis*. The author covers the cases of a large number of mathematicians who left Germany during the time when the Nazi party was in power, documenting the circumstances that led to their decisions to emigrate and the support networks that, in some cases, eased the difficulties of their transitions.
 Most of us know that many academics fled Germany due to persecution of intellectuals, especially Jewish scholars, and that the majority immigrated to the United States. This book took me well beyond that basic understanding; it discusses mathematicians previously unknown to me, like those who fled to such destinations as Turkey and Ecuador. Despite the fact many of these emigrants had much to offer to the academic communities in their new countries and a number of those who came to the U.S.A. made crucial contributions to the war effort, they were hardly welcomed with open arms. They frequently encountered anti-Semitism or hostility from citizens of their host countries who regarded them as competitors for available academic positions. Many potential countries of refuge imposed quotas on the number of entry visas available. Academic institutions willing to hire German mathematicians could choose from many who were available and those judged to be lesser talents or past their prime were often eliminated from consideration.
 Adjustment to the customs and academic culture of a new country was problematic for many of

these immigrants. Those who were unable to secure permanent academic appointments sometimes encountered extreme financial difficulties. This book is a reminder that mathematics is the pursuit of real people whose lives are inevitably influenced by the surrounding culture and events. Each person's story is unique and this work gives insight into the experiences of scores of mathematicians who lived through this difficult period of history. I enjoyed this book, learned a lot from it and heartily recommend it to those who are interested in the history of mathematics.

Lamarr Widmer

The Proof is in the Pudding, by Steven G. Krantz, New York, Springer, 2010, 264 pp. $39.95 (hardcover), ISBN 978-0-387-48908-7

In Chapter 1, Krantz explains what a proof is and what purpose it serves in the discipline of mathematics. Along the way he talks about Platonism and Kantianism, the experimental nature of mathematics, the role of conjectures and the dissemination of mathematics. Chapters 2 through 4 are a chronological survey of the development of the concept of mathematical truth up to the 20th century. The much longer Chapter 5 covers 20th century developments and players, including Hilbert, Bourbaki, Erdös and Halmos. Rather than just proofs, this book is about how mathematicians work. On page 76, Krantz says, "Mathematics is actually a *process*, and that process does not always divide itself neatly into theorems and proofs."

Chapter 7 deals with computer-generated proofs and Chapter 8 considers the computer as a teaching tool and potential substitute for proof. At this point, the book becomes much more thought provoking. In the final section of chapter 8, titled "Mathematical Communication," Krantz discusses the changes brought about by electronic communication, including the posting of lecture notes and preprints, electronic archiving, the use of TEX and bypassing of editors and referees. He says that many electronic journals are a "free-for-all" (p. 147) and discusses the consequences of this situation.

Chapter 9, "Aspects of Modern Mathematical Life," considers the role of universities and mathematical institutes in the mathematical community. Here we find mention of the Clay Institute's Millennium Problems and of the IBM Selectric Typewriter. Chapters 10-12 include material on the classification of finite simple groups, the Kepler sphere-packing problem, Thurston's geometrization program, Perelman and the Poincaré Conjecture, the Riemann Hypothesis, and Wiles' proof of Fermat's Last Theorem. Chapters 12 and 13 briefly look to the future, raising questions and offering cogent speculations.

This work includes an index of names and 9 pages of references. The author does an admirable job of illuminating a topic of great significance and depth. I found much material of interest in these pages and a second reading would no doubt reveal even more. As a side note, the author includes several quotes at the beginning of each chapter. This has replenished my supply of catchy quotes to use at the end of future exam papers.

Lamarr Widmer

Perfect Rigor, by Masha Gessen, New York, Houghton Mifflin Harcourt Publishing Company, 2009, 242 pp. $26.00 (hardcover), ISBN 978-0-15-101406-4

After reading the excellent article "A Conversation with Masha Gessen" (*Math Horizons, 17*:2, November 2009), I could not wait to get my hands on this book. Now I have done so and it certainly does not disappoint. This work documents Grigory Perelman's proof of the Poincaré Conjecture, the pure mathematical achievement of the current decade, just as Andrew Wiles' proof of Fermat's Last Theorem highlighted the 1990's. The statement of the Poincaré Conjecture is considerably more difficult to understand than Fermat's Last Theorem. In both cases, understanding of the proof requires great expertise in a particular mathematical specialty. Interest in the Poincaré Conjecture was heightened in 2000 when the Clay Mathematics Institute offered $1 million to the author of the first correct proof.

Gessen discusses the process of making Perelman's proof known to the mathematical world, checking the details and reaching consensus as to its correctness. Perelman spent seven years composing his proof and first made it known by posting a preprint on the website arXiv.org. This paper caused a sensation among topologists working in the relevant specialty, many of whom had never heard of Perelman. Gessen reports many topologists apparently stayed up all night reading it and by the following day, e-mails were flying back and forth between them. This was the beginning of the process that resulted in the acceptance of his proof. Gessen speculated the $1 million would eventually be offered to Perelman and that he would refuse it. We now know that she was right on both counts.

This leads us to the personality of Perelman himself, easily the most interesting aspect of this story. Perelman is the product of a rigorous Russian program aimed at identifying and developing young mathematical talent while preparing the very best for competitive mathematical problem-solving, particularly the International Mathematical Olympiad. He distinguished himself academically, overcoming barriers erected to limit enrollment of Jews. His academic career took him to a number of Russian and American educational institutions where his genius was recognized and his eccentric character noted. By the time his proof was verified he had cut himself off from former colleagues and teachers. He was greatly offended by the publicity and job offers that resulted from his achievement. He resented any insinuation that the million dollar prize played any part in motivating his accomplishment. He refused all interview requests. In chapter 10, Gessen makes a convincing case for the theory that Perelman has Asperger's syndrome, a form of autism.

Masha Gessen is uniquely suited for the task of writing such a work. She is herself a product of the Russian math whiz program. She was able to overcome Perelman's refusal to speak by tracking down and interviewing many of his former fellow students, coaches and colleagues and interviewing them. She is able to give a competent and understandable summary of the meaning and significance of the Poincaré Conjecture. Finally, she has produced an excellent work in English. We are fortunate to have such a qualified author to produce the first book documenting this episode in the ongoing mathematical story.

Lamarr Widmer

Numbers Rule, by George G. Szpiro, Princeton, New Jersey, Princeton University Press, 2010, 226 pp. $26.95 (hardcover), ISBN 978-0-691-13994-4

The mathematics of voting procedures is now a somewhat common topic in the undergraduate university curriculum. I have taught it myself, using a textbook that focused mostly on the details of different methods and the inherent paradoxes of the topic. Due to this experience, I acquired knowledge of the history of this topic in the American political context. But I learned a lot more from Szpiro, who goes far beyond the scope of the typical textbook chapter covering this topic.

The greatest contribution of this work is its comprehensive coverage of the history of this topic. In chapter 1 we learn of Plato's interest in the subject and his extensive recommendations for voting procedures in the Greek democracy. Szipiro highlights the strikingly *un*democratic nature of Plato's preferred election processes. Chapter 2 covers the similar interest and involvement of Pliny the Younger in political affairs of the Roman Empire. Chapter 3 covers the writing of thirteenth century scholar, Ramon Llull, discovered recently among manuscripts in the Vatican library. As a theologian, he was particularly interested in church elections and proposed a method for choosing a winner from among several candidates by means of a series of two-by-two matchups. At this point, Szpiro introduces the principle of *transitivity*, which says that if A beats B and B beats C, then A beats C. Llull's method does not possess this desirable property. This is an example of how Szipiro's timely introduction of the mathematical concepts weaves them into his historical account.

In this fashion, Szpiro brings us, chapter by chapter, to the present, along the way introducing more familiar figures such as Condorcet, de Borda and Kenneth Arrow, along with their contributions. While he does dwell on the fascinating and contentious question of apportionment of the U. S. House of Representatives, Szipiro's presentation is not American-centric, as are the textbooks mentioned above. He also gives appropriate attention to the issues of allocation of seats in the Swiss Parliament and Israel's Knesset. This is not a textbook; there are no exercises. It does offer a four-page bibliography as well as appendices of additional reading (in particular, biographical sketches of key players) for each chapter. This is an excellent work on a topic that focuses the penetrating light of mathematical reasoning on issues of the humanities that inescapably touch our lives i.e. decision-making, politics, power and ethics.

Lamarr Widmer

Mythematics, by Michael Huber, Princeton, NJ, Princeton University Press, 2009, 183 pp. $24.95 (hardcover), ISBN 978-0-691-13575-5

The title of this book refers to Greek mythology, particularly the twelve labors of Hercules. In his introduction, Huber quotes classical Greek author Apollodorus, who explains that by completing these tasks Hercules would achieve immortality. The tasks are assigned to Hercules by Eurystheus, who does not expect him to succeed in completing them. Nine of the tasks involve vanquishing some mythical creature. Each of Huber's chapters begins with a passage from Apollodorus, describing one of the twelve labors.

These myths of Hercules are the foundation on which Huber builds his mathematical tale. For each of the twelve labors, he proposes a number of related mathematical problems some of which, when solved, point the way to Hercules' success. Chapter 8 is typical. The eighth labor is called The Horses of Diomedes. Hercules must capture the man-eating mares belonging to Diomedes and deliver them to Eurystheus. Diomedes is king of the Bistones who pursue Hercules when his theft is discovered. In the course of Hercules' escape, the horses kill his young ally, Abderus. Hercules founds a city named Abdera, in his honor.

Working from this myth, Huber imagines three associated math problems. The first is a pursuit problem, with Hercules fleeing directly south from the horse stables and the Bistones first going west from their quarters to the stables and then turning south in pursuit of Hercules. Its solution requires modeling the trajectories of Hercules and the Bistones and using differentiation to determine the closing distance between them. The second problem concerns the trajectory of a stone launched from Hercules' sling as he battles the Bistones and requires differentiation to find both velocity and acceleration. The third problem considers the population growth of the city Abdera, which is modeled by a logistic curve. The solution uses a differential equation.

Many of Huber's problems are reworks of familiar textbook problems or of classics such as the Josephus problem. In all cases he has cleverly connected them to the Hercules narrative. He provides a solution for every problem, at times leaving details to be checked by the reader. In an appendix, he lists all problems by subject area. The list is impressive, including algebra, calculus, combinatorics, difference equations, differential equations, geometry, probability and trigonometry. While Greek mythology is not my cup of tea, I can appreciate Huber's imaginative approach to bringing mathematics into this setting.

Lamarr Widmer

ALPHAMETICS

Contributed by Paul Boymel

Since this book is dedicated to the memory of Joseph S. Madachy, it is only fitting that this section opens with two items authored by him. The first is a short history of alphametics that appeared in issue number 6 of **Recreational Mathematics Magazine** and the second is the first alphametic authored by Joe that was published in **Journal of Recreational Mathematics**. It appeared in volume 1 number 2.

Alphametics (Recreational Mathematics Volume 6, 1961.)
by Joseph S. Madachy

The interest of RMM readers in this department prompts the Editor to present a bit of history of the type of puzzles found here.

The very first "hidden digit" problem will probably never be known, but it is easy enough to go back to 1906 when W. E. H. Berwick presented a division problem of 79 digits with only seven sevens given as clues. (**School World** Vol. VIII pp. 280, 320). Prof. Schuh of Delft in 1921 managed to devise a long division problem in which *none* of the digits were given directly as clues. These "hidden digit" problems did not utilize letters or symbols to match particular digits – blank spaces were left to be filled in. **Strand Magazine**, in 1921, published a "hidden digit" problem using letters of the alphabet and, later, some problems with chessmen covering the digits appeared. The word *cryptarithmie* (cryptarithmetic), attributed to Minos, was used in the May 1931 issue of **Sphinx**.

So far, any letters or symbols could be used in place of the digits. However, letter substitutions which formed words and even phrases proved more popular. A truly ideal cryptarithm is one in which all the digits from 0 through 9 are substituted by letters which form a sensible phrase and in which only *one* solution is possible.

J. A. H. Hunter, who devised the term Alphametic to apply to a cryptarithm in which the letter substitutions make sense, first used the term *alphametic* in the **Toronto Globe & Mail** of October 27, 1955. The term is credited to a typographical error made by one of Mr. Hunter's correspondents at that time.

Here it is, the first cryptarithm published under the name *Alphametic*. We leave it to be solved by **RMM** readers.

```
BE ) ABLE ( SIR
     MR
     ___
     RRL
     RLM
     ___
      BE
      BE
```

Newspaper Alphametic (**Journal of Recreational Mathematics** Volume 1, No. 2, 1968. (Joseph S. Madachy)

```
    DANDY        The Record-Courier is the daily newspaper in
   RECORD        Kent and Ravenna, Ohio
   _____
  COURIER
```

It is with deep regret that I also announce the passing of longtime **Journal of Recreational Mathematics** problem contributor and solver Paul Boymel. He will be missed and the remainder of this section is dedicated to his memory.

1. United Nations – 1 by Paul Boymel, Potomac, Maryland

```
        CHINA x TOGO  =  MONGOLIA
```

2. United Nations – 2 by Paul Boymel, Potomac, Maryland

```
        MALTA x CHAD  =  TANZANIA
```

3. United Nations – 3 by Paul Boymel, Potomac, Maryland

```
        SPAIN x LAOS  =  SANMARINO
```

4. United Nations – 4 by Paul Boymel, Potomac, Maryland

```
        BENIN x OMAN  =  CAMEROON
```

5. United Nations – 5 by Paul Boymel, Potomac, Maryland

```
        INDIA x IRAN  =  DJIBOUTI
```

6. United Nations – 6 by Paul Boymel, Potomac, Maryland

```
SYRIA x IRAQ  =  SLOVAKIA
```

PROBLEMS AND CONJECTURES
contributed by Lamarr Widmer

Since this book is dedicated to the memory of Joseph S. Madachy, it is only fitting that we start with a reprinting of an old problem that was created by Joe and references Martin Gardner. This problem appeared in **Recreational Mathematics Magazine Issue No. 6**, December 1961.

Related to a puzzle posed by Martin Gardner in "Scientific American" (June 1960, page 168) is the following: What is the shortest strip of paper one inch wide, and black on one side, which can be folded to form a cube which is black on all six sides?

1. Mighty Magic by Clarence Gipbsin, Represa, CA

How many magic squares are present in Figure 1, what are their orders and how are they embedded in the 27 by 27 square?

2. Pentomino Words II by Brian Barwell, Hampton, Middlesex, UK

Figure 2 shows the twelve pentominoes and their associated reference letters. A pentomino word is one containing only the letters F, I, L, N, P, T, U, V, W, X, Y and Z such that the pentominoes corresponding to the letters in the word can be arranged to form a rectangle. Repeated letters are allowed. Figure 3 shows an example of such a rectangle made from the letters of the six-letter word PULPIT.

Find a pentomino word containing more than six letters. What is the longest pentomino word?

3. Comeback 2511 by Paul E. Boymel, Potomac, MD

Problem 2511 (*JRM 30*:2, p. 148, *JRM 31*:2, p. 145) asked for an arrangement of the integers 1 through 25 in a square array so that each number except 1 and 2 is the sum of two of its neighbors, orthogonal or diagonal. This problem has been found to have roughly 5000 solutions.

a. Find a solution where the number 1 is on the outside edge of the square.
b. Find the solution having the largest possible sum of any row, column or diagonal.
c. Find the solution having the smallest possible sum of any row, column or diagonal.

Figure 1

638	71	476	631	64	469	636	69	474	575	8	413	568	1	406	573	6	411	620	53	458	613	46	451	618	51	456
233	395	557	226	388	550	231	393	555	170	332	494	163	325	487	168	330	492	215	377	539	208	370	532	213	375	537
314	719	152	307	712	145	312	717	150	251	656	89	244	649	82	249	654	87	296	701	134	289	694	127	294	699	132
633	66	471	635	68	473	637	70	475	570	3	408	572	5	410	574	7	412	615	48	453	617	50	455	619	52	457
228	390	552	230	392	554	232	394	556	165	327	489	167	329	491	169	331	493	210	372	534	212	374	536	214	376	538
309	714	147	311	716	149	313	718	151	246	651	84	248	653	86	250	655	88	291	696	129	293	698	131	295	700	133
634	67	472	639	72	477	632	65	470	571	4	409	576	9	414	569	2	407	616	49	454	621	54	459	614	47	452
229	391	553	234	396	558	227	389	551	166	328	490	171	333	495	164	326	488	211	373	535	216	378	540	209	371	533
310	715	148	315	720	153	308	713	146	247	652	85	252	657	90	245	650	83	292	697	130	297	702	135	290	695	128
593	26	431	586	19	424	591	24	429	611	44	449	604	37	442	609	42	447	629	62	467	622	55	460	627	60	465
188	350	512	181	343	505	186	348	510	206	368	530	199	361	523	204	366	528	224	386	548	217	379	541	222	384	546
269	674	107	262	667	100	267	672	105	287	692	125	280	685	118	285	690	123	305	710	143	298	703	136	303	708	141
588	21	426	590	23	428	592	25	430	606	39	444	608	41	446	610	43	448	624	57	462	626	59	464	628	61	466
183	345	507	185	347	509	187	349	511	201	363	525	203	365	527	205	367	529	219	381	543	221	383	545	223	385	547
264	669	102	266	671	104	268	673	106	282	687	120	284	689	122	286	691	124	300	705	138	302	707	140	304	709	142
589	22	427	594	27	432	587	20	425	607	40	445	612	45	450	605	38	443	625	58	463	630	63	468	623	56	461
184	346	508	189	351	513	182	344	506	202	364	526	207	369	531	200	362	524	220	382	544	225	387	549	218	380	542
265	670	103	270	675	108	263	668	101	283	688	121	288	693	126	281	686	119	301	706	139	306	711	144	299	704	137
602	35	440	595	28	433	600	33	438	647	80	485	640	73	478	645	78	483	584	17	422	577	10	415	582	15	420
197	359	521	190	352	514	195	357	519	242	404	566	235	397	559	240	402	564	179	341	503	172	334	496	177	339	501
278	683	116	271	676	109	276	681	114	323	728	161	316	721	154	321	726	159	260	665	98	253	658	91	258	663	96
597	30	435	599	32	437	601	34	439	642	75	480	644	77	482	646	79	484	579	12	417	581	14	419	583	16	421
192	354	516	194	356	518	196	358	520	237	399	561	239	401	563	241	403	565	174	336	498	176	338	500	178	340	502
273	678	111	275	680	113	277	682	115	318	723	156	320	725	158	322	727	160	255	660	93	257	662	95	259	664	97
598	31	436	603	36	441	596	29	434	643	76	481	648	81	486	641	74	479	580	13	418	585	18	423	578	11	416
193	355	517	198	360	522	191	353	515	238	400	562	243	405	567	236	398	560	175	337	499	180	342	504	173	335	497
274	679	112	279	684	117	272	677	110	319	724	157	324	729	162	317	722	155	256	661	94	261	666	99	254	659	92

Figure 2

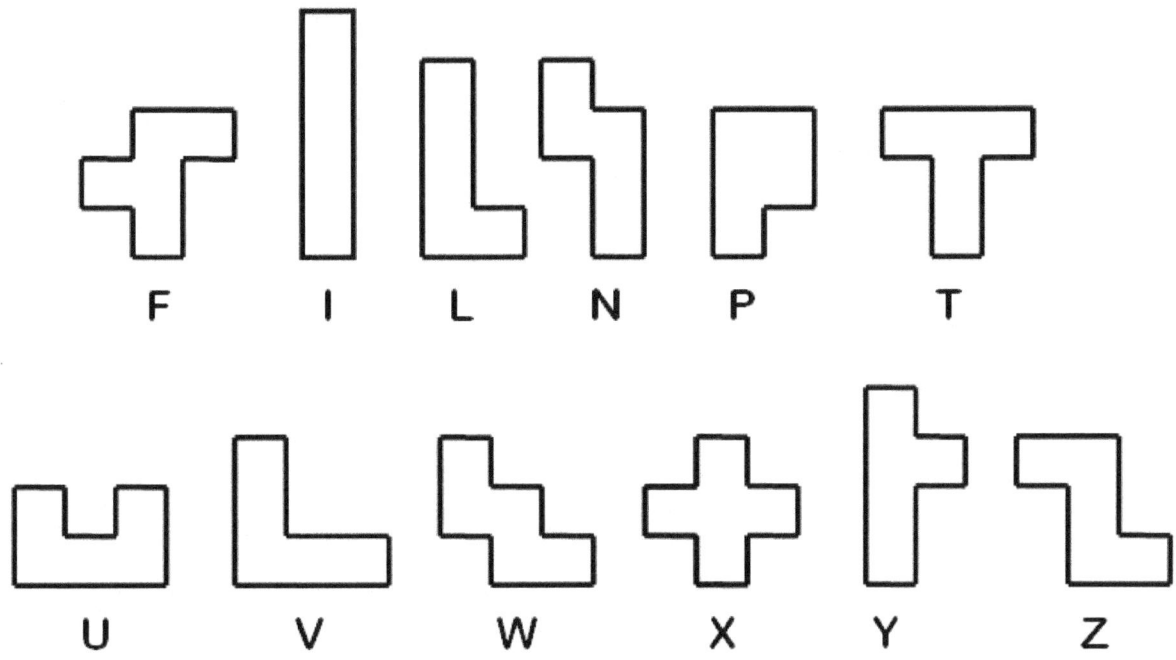

F I L N P T

U V W X Y Z

Figure 3

PULPIT

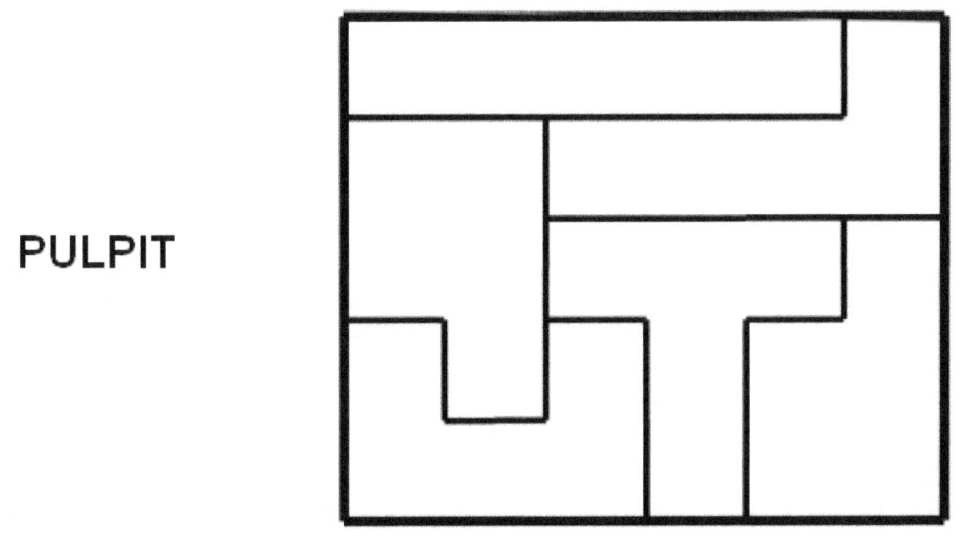

SOLUTIONS TO PROBLEMS AND CONJECTURES FROM JOURNAL OF RECREATIONAL MATHEMATICS 37(3)

Edited by Lamarr Widmer

This section contains solutions to problems that appeared in **Journal of Recreational Mathematics Volume 37 No. 3**. The editor is grateful to Sage Publications for kindly giving permission for this material to be published here.

2850. A Thirteen-Coin Weighing Problem by Robert Khanis, Sydney, Australia (*JRM* 37:3, p. 255)

You are given a set of thirteen coins. Their weights in grams are stated to be 2, 3, 5, 7, 11, 13, 17, 19, 23, 29, 31, 37, and 41. You know that, in fact, one of them varies from its stated weight by a detectable amount which is no more than 0.5 gram. How can you identify the defective coin with three weighings on a balance scale and determine whether it is heavier or lighter than its stated weight?

Solution by Brian Barwell

Separate the weights into three sets as follows: $= \{2, 3, 13, 29, 31\}$, $B = \{5, 17, 19, 37\}$ and $C = \{7, 11, 23, 41\}$. First weigh A against B. There are three possible outcomes: (1) they balance, (2) A is heavier, (3) A is lighter. We consider these cases separately.

(1) In this case the bad coin is in C. For the second weighing, we test $\{7, 23, 41\}$ against $\{5,29,37\}$. If they balance, the 11-gram coin is bad and weighing $\{2,11\}$ against $\{13\}$ will show whether it is heavy or light. If they do not balance, the bad coin is 7, 23 or 41 and the result will show whether it is heavy or light. Then weighing $\{3,7,11\}$ against $\{23\}$ will identify the bad coin.

(2) In this case we know that either A contains a heavy coin or B has a light one. For the second weighing, test $\{2,3,29,5,17,19\}$ against $\{11,23,41\}$. If they balance, the bad coin is a heavy 13, a heavy 31 or a light 37. Then for the third weighing, test $\{13,37\}$ against $\{2,7,41\}$. If they balance, 31 is heavy; otherwise the result will determine if 13 (heavy) or 37 (light) is the bad coin. If the second weighing does not balance, we will be able to determine whether the bad coin is (a) a light 5, 17 or 19, or (b) a heavy 2, 3 or 29. Then the bad coin can be identified by the following for the third weighing: (a) weigh $\{2,17\}$ against $\{19\}$ or (b) weigh $\{3,7,19\}$ against $\{29\}$.

(3) If A is lighter than B, then proceed in (2) but interchange the words "heavy" and "light".

2851. Pyramid Perplexity by Henry Ibstedt, Issy les Moulineaux, France (*JRM* 37:3, p. 256)
Calculate the altitude of a quadrilateral pyramid whose base diagonals intersect at right angles and where the foot of the altitude is at the intersection of the base diagonals. The areas of the triangular sides are *a, b, c* and *d.*

Solution by Daniele Degiorgi

Let the origin $O = (0,0,0)$ coincide with the intersection of the base diagonals. Then we can assume for the vertices the coordinates $A = (x_1,0,0)$, $B = (0,y_1,0)$, $C = (x_2,0,0)$, $D = (0,y_2,0)$, $H = (0,0,h)$. It follows that the edges of the base have lengths

$$\|AB\| = \sqrt{x_1^2 + y_1^2} \, , \|BC\| = \sqrt{y_1^2 + x_2^2} \, , \|CD\| = \sqrt{x_2^2 + y_2^2} \, , \|DA\| = \sqrt{y_2^2 + x_1^2} \, .$$

The base triangles AOB, BOC, COD, DOA have respectively the areas

$$\frac{x_1 y_1}{2} = \frac{h_1 \|AB\|}{2} \, , \frac{y_1 x_2}{2} = \frac{h_2 \|BC\|}{2} \, , \frac{x_2 y_2}{2} = \frac{h_3 \|CD\|}{2} \, , \frac{y_2 x_1}{2} = \frac{h_4 \|DA\|}{2}$$

where h_1, h_2, h_3 and h_4 are the perpendicular distances from O to the sides of $ABCD$.

Then the areas of the triangular faces are

$$a = \frac{\|AB\| \sqrt{h_1^2 + h^2}}{2} \, , \quad b = \frac{\|BC\| \sqrt{h_2^2 + h^2}}{2} \, , \quad c = \frac{\|CD\| \sqrt{h_3^2 + h^2}}{2} \, , \quad d = \frac{\|DA\| \sqrt{h_4^2 + h^2}}{2} \, .$$

And $h_1 = \frac{x_1 y_1}{\|AB\|}$, $h_2 = \frac{y_1 x_2}{\|BC\|}$, $h_3 = \frac{x_2 y_2}{\|CD\|}$, $h_4 = \frac{y_2 x_1}{\|DA\|}$.

Squaring the areas, we have

$$\begin{cases} 4a^2 = x_1^2 y_1^2 + (x_1^2 + y_1^2)h^2 \\ 4b^2 = y_1^2 x_2^2 + (y_1^2 + x_2^2)h^2 \\ 4c^2 = x_2^2 y_2^2 + (x_2^2 + y_2^2)h^2 \\ 4d^2 = y_2^2 x_1^2 + (y_2^2 + x_1^2)h^2 \end{cases}$$

Now we easily find that

$$16a^2c^2 - 16b^2d^2 = h^4\left(x_1^2y_2^2 + y_1^2x_2^2 - x_2^2y_2^2 - y_1^2x_1^2\right)$$

$$4b^2 + 4d^2 - 4a^2 - 4c^2 = x_1^2y_2^2 + y_1^2x_2^2 - x_2^2y_2^2 - y_1^2x_1^2$$

$$h = \sqrt{2}\sqrt[4]{\frac{a^2c^2 - b^2d^2}{b^2 + d^2 - c^2 - a^2}}$$

2852. Sibling Primes by Hubert Hagadorn, Menlo Park, CA (*JRM* 37:3, p. 256)

Expressing the number 13 in bases 2 through 9, we have 1101_2, 111_3, 31_4, 23_5, 21_6, 16_7, 15_8, 14_9. The siblings of 13 are the numbers named by these same eight sequences of digits considered as base 10 numerals. In this case we find that only 31 and 23 are prime. Find a prime number whose eight siblings are all prime.

Solution by Ken Klinger

This problem is addressed, in part, on the prime puzzle website at http://primepuzzles.net/puzzles/puzz_024.htm . Four primes which meet the conditions of this problem, are known. They are 50,006,393,431; 727,533,146,383; 2,250,332,130,313 and 2,651,541,199,513.

2853. Pentasquare Knight's Tour by Donald E. Knuth, Stanford, California (*JRM* 37:3, p. 256)

Figure 1 shows the *pentasquare,* an arrangement of 30 distorted squares in a symmetrical pattern, where every internal vertex is the meeting point of five "squares". A *knight move* in the pentasquare is analogous to a knight move on an ordinary chessboard, except that we imagine that the squares have not been distorted. "Either move one step, turn, and then two steps; or move two steps, turn, and then one step." Figure 1 shows a cycle of four knight moves that return to the starting point. It is not difficult to see that there are no 3-cycles, but 5-cycles are possible. Find a *knight's tour* of the pentasquare i.e. a 30-cycle in which every "square" is visited exactly once. Try to make your solutions as symmetrical as possible.

Figure 1

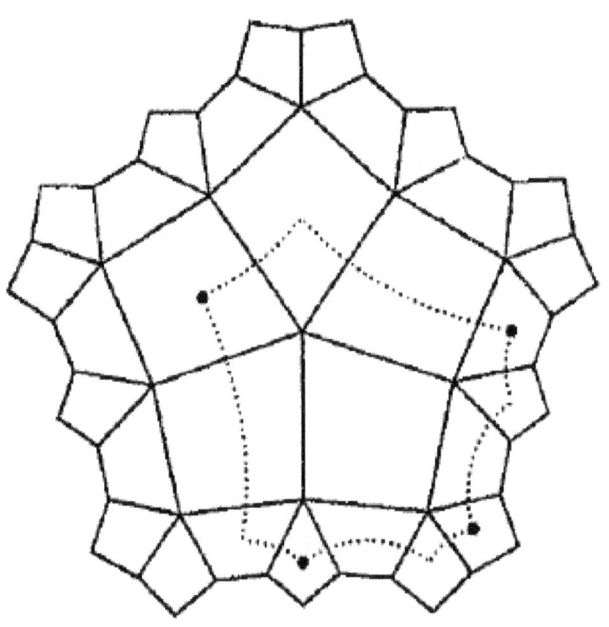

Solution by Daniele Degiorgi

Of the 22456 tours the one in figure 2 seems to be the most symmetrical

2854. JRM Polyomino Rectangle by Dario Uri, Pontecchio Marconi, Italy (*JRM* 37:3, p. 256)

Assemble the twelve pentominoes (Figure 3) and the pieces *J*, *R* and *M* shown in Figure 4 to form a 9×10 rectangle. Is there a solution in which the *J, R, M* pieces do not touch each other?

Solution by Lionel Bidwell

See Figure 5.

Figure 2

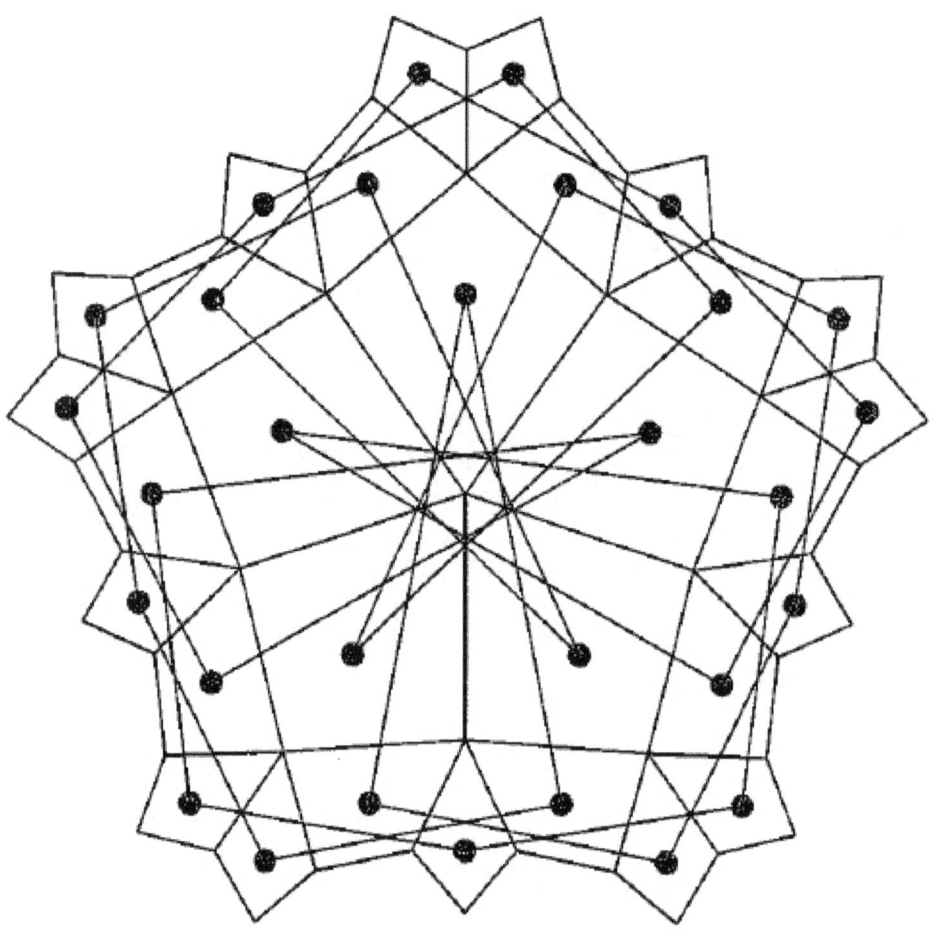

Figure 3

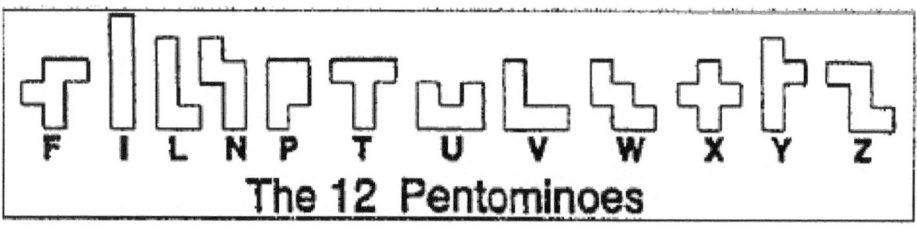

The 12 Pentominoes

Figure 4

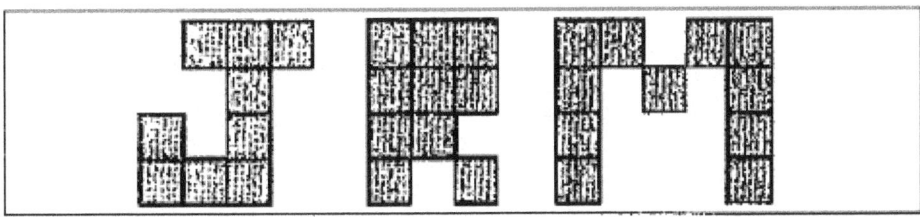

Figure 5

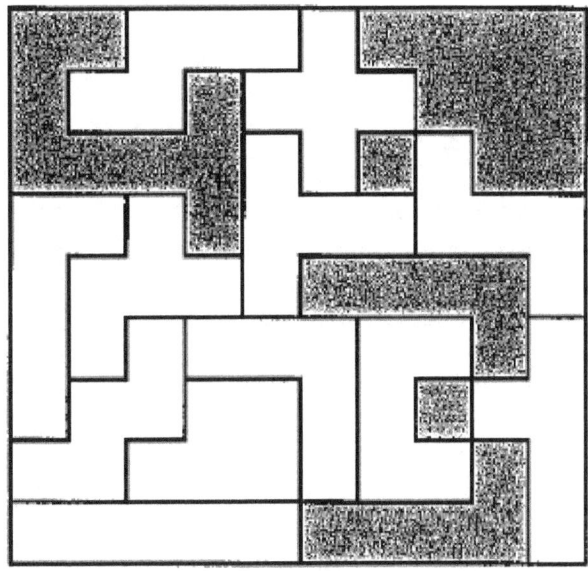

2855. Tangent Circles by Ken Klinger, Northbrook, Illinois (*JRM* 37:3, p. 256)

In Figure 6, a circle of radius a is inscribed in a circle of radius $2a$ so that it is tangent to the larger circle at the point $(0,2a)$. A circle with center B is tangent to the larger and smaller circle as well as the x-axis. A circle with center C is tangent to the circle centered at B and circle centered at $(0,a)$ as well as to the x-axis. What are the coordinates of the points B and C? This problem was inspired by a wooden decoration at the front of some meeting rooms in the Sheraton Hotel and Towers in Chicago.

Solution by Brian Barwell

We denote the radii of the circles centered at B and C by b and c, respectively and their centers by (x_B, b) and (x_C, c). AB has length $a + b$, so by the Pythagorean Theorem,

$$(a+b)^2 = x_B^2 + (a-b)^2 \text{ or } x_B^2 = 4ab. \text{ Similarly, we find } x_C^2 = 4ac \text{ and}$$

$$(x_B - x_C)^2 = 4bc.$$

Also, since the length of OB is $2a - b$, we have $(2a - b)^2 = x_B^2 + b^2$ or $x_B^2 = 4a^2 - 4ab$.

From $x_B^2 = 4ab = 4a^2 - 4ab$, we get $x_B^2 = 2a^2$ and then $x_B = -a\sqrt{2}$. Then from the equation $x_B^2 = 4ab$, we have $b = \frac{a}{2}$. From equations $x_C^2 = 4ac$ and $(x_B - x_C)^2 = 4bc$, we get quadratic equations for c and x_c. These equations have two solutions,
$(x_c, c) = \left(-2(\sqrt{2} - 1)a, (3 - 2\sqrt{2})a\right)$ and $(x_c, c) = \left(-2(\sqrt{2} + 1)a, (3 + 2\sqrt{2})a\right)$. The first of these solves our problem and the second applies to the case where the circle C lies above the line AB.

2856. Integer-Valued Functions by Lamarr Widmer, Grantham, Pennsylvania (JRM 37:3, p. 258)

Let S be the sum-of-digits function, e.g. $S(851) = 8 + 5 + 1 = 14$, and let $T(n)$ denote the nth triangular number, e.g. $T(4) = 1 + 2 + 3 + 4 = 10$. Define
$F(n) = S^k\big(T(n)\big) = S(S \dots \big(S(T(n))\big)\dots)$, where k is sufficiently large to produce a single-digit result. Find, with proof, the range of $F(n)$.

Solution by Ken Klinger

$F(n)$ computes what is known as the digital root of the number $T(n)$. The digital root of a number k is just $k \pmod 9$ except when $k \pmod 9 = 0$, in which case $S(k) = 9$. It is well known that $T(n) = \frac{n(n+1)}{2}$. Checking all nine possible values of $T(n)$ modulo 9, we find that $T(n)$ can only equal 1, 3, 6 or 0 modulo 9. Therefore, the range of $F(n)$ is $\{1, 3, 6, 9\}$.

*2857. A Variation on Fortune's Conjecture by Andrew Cusumano, Great Neck New York (JRM 37:3, p. 259)

Let p_k denote the kth prime number and let $N(n)$ be the next-prime function, i.e. $N(n)$ is the least prime which is greater than n. Finally, define

$$G(k) = N\big((2 \times 3 \times 5 \times \ldots \times p_{k-1}) + p_k\big) - (2 \times 3 \times 5 \times \ldots \times p_{k-1})$$

Then, for example, $G(4) = N\big((2 \times 3 \times 5) + 7\big) - (2 \times 3 \times 5) = 41 - 30 = 11$.

Is every value of $G(k)$ a prime?

Solution by Daniele Degiorgi

As the meaning of $G(1)$ is open to interpretation, we assume that $k > 1$. Following Guy [1], we let $p_k\#$ denote the product $2 \times 3 \times 5 \times \ldots \times p_{k-1} \times p_k$. Then
$G(k) = N(p_{k-1}\# + p_k) - p_{k-1}\#$. Now we let $N(p_{k-1}\# + p_k) = p_{k-1}\# + p_k + r_k$ where
$r_k > 0$ and all numbers $p_{k-1}\# + p_k + j$, for $0 < j < r_k$ are composite. It follows that
$G(k) = p_k + r_k$. We note that r_k must be even, for otherwise $p_{k-1}\# + p_k + r_k$ is a multiple of
2 and therefore not prime.
Suppose now that $G(k)$ is not prime. Then we would have that $p_{k-1}\# + p_k + r_k$ is prime and
$p_k + r_k = pq$ with $p > 1, q > 1$. Neither p nor q can be divisible by any of the p_j for $j < k$.
Thus p and q have only prime divisors greater than p_k. It follows that $r_k \geq p_k(p_k - 1)$. This
implies that there is a difference greater than r_k between the prime $p_{k-1}\# + p_k + r_k$ and the
previous prime. This would seem to be a counterexample to Schnitzel's conjecture (p.1 of [1]),
that for $x > 7$, there is always a prime between x and $x + (\ln x)^2$.
Anyway, I was able to verify the conjecture up to $k = 1000$.

Reference

1. Richard K. Guy, *Unsolved Problems in Number Theory*, 3e., Problem Books in Mathematics, Springer-Verlag, New York, 2004.

*2858. Enterprise Numbers by Fred Barnes, Thayer, Missouri (*JRM* 37:3, p. 259)

Let x, y and n be positive integers with $n \geq 3$. If $E = x^n + y^n - p = (x + y)^n - q$ where p
and q are perfect numbers, then E is an *enterprise number*. For example,

$E = 1701 = 12^3 + 1^3 - 28 = (12 + 1)^3 - 496$. Find another enterprise number or prove that
none exist.

Solution by Richard I. Hess

I did not find any other enterprise numbers, using the perfect numbers $2^{p-1}(2^p - 1)$ for values of p up to 19. Proving impossibility seems out of the question as long as there are an unknown number of undiscovered odd perfect numbers.

2859. Diophantine System by John Wahl, Mt. Pocono, Pennsylvania (*JRM* 37:3, p. 259)

Find a solution for the following system in which the twelve variables equal distinct positive integers.

$$\begin{cases} abw + ezd + gxd - (ady + efw + ghw) = 0 \\ azb + cfw + chy - (cxb + efz + ghz) = 0 \\ cbw + efy + ghy - (dyc + ezb + bgx) = 0 \\ adz + efx + ghx - (dcx + fwa + hya) = 0 \end{cases}$$

This problem was submitted by Andrew Cusumano who states that his colleague, Mr. Wahl is deceased.

Solution by Brian Barwell

The four equations can be written as

$$\begin{aligned} a(bw - dy) + e(dz - fw) + g(dx - hw) &= 0 \\ b(az - cx) + f(cw - ez) + h(cy - gz) &= 0 \\ c(bw - dy) + e(fy - bz) + g(hy - bx) &= 0 \\ d(az - cx) + f(ex - aw) + h(gx - ay) &= 0 \end{aligned}$$

We can satisfy these equations by making each term in parentheses equal to zero. This will be accomplished if $\frac{w}{e} = \frac{x}{a} = \frac{y}{g} = \frac{z}{c}$ and $\frac{w}{d} = \frac{x}{h} = \frac{y}{b} = \frac{z}{f}$. To make all of these integers distinct, we can choose $(w, x, y, z) = (3,4,5,7)$ and then $\frac{3}{6} = \frac{4}{8} = \frac{5}{10} = \frac{7}{14}$ and $\frac{3}{9} = \frac{4}{12} = \frac{5}{15} = \frac{7}{21}$. So our solution is $(a,b,c,d,e,f,g,h,w,x,y,z) = (8,15,14,9,6,21,10,12,3,4,5,7)$.

Editor's Commentary
Daniele Degiorgi, using a different method of making the values distinct, found $(a,b,c,d,e,f,g,h,w,x,y,z) = (1,55,2,77,7,22,5,11,21,3,15,6)$.
Henry Ibstedt found $(a,b,c,d,e,f,g,h,w,x,y,z) = (9,12,7,6,11,10,5,8,2,4,3,1)$. In this solution, none of Barwell's twelve terms is zero.
Richard Hess reports finding 300 solutions using just the numbers 1 through 12.

SOLUTIONS TO ALPHAMETICS APPEARING IN THIS ISSUE

1. 58702 x 1636 = 96036472

2. 50370 x 1408 = 70920960

3. 26514 x 9582 = 254057148

4. 12565 x 7435 = 93420775

5. 28624 x 2348 = 67209152

6. 95012 x 1026 = 97482312

SOLUTIONS TO ALPHAMETICS THAT APPEARED IN JOURNAL OF RECREATIONAL MATHEMATICS VOLUME 37(3)

Contributed by Steven Kahan

Once again I would like to acknowledge that Sage Publications was kind enough to give permission for these problems to be re-published here.

2837. No Freedom of Speech by Chuck Porter, Freeport, Texas

```
FIRE + FIRE +FALSE = ALARM
```

Sound the greatest ALARM here.

Solution:

```
5483 + 5483 + 56723 = 67689
```

2838. Simply Sum-Thing by Alf D. Seider, Bayside, New York

```
ONE + ONE + ONE + EIGHT + NINE + THIRTY + THIRTY = EIGHTY
```

```
983 + 983 + 983 + 36051 + 8683 + 156417 + 156417 = 360517
```

2839. Annual Alphametic -1 by Junya Take, Kanagawa, Japan

```
VI + VII + XVIII + XVIII + CMLXI + MIII = MMXIII
```

```
56 + 566 + 45666 + 45666 + 21046 + 1666 = 114666
```

2840. Annual Alphametic – 2 by Junya Take, Kanagawa, Japan

```
VIII + XVII + XXIII + CMLXX + CMXCV = MMXIII
```

```
1888 + 3188 + 33888 + 92533 + 92391 = 223888
```

2841. 800 in Italiano – Uno by Giulio Cesare, Rome, Italy

```
NOVE + UNDICI + VENTUNO + VENTITRE + TRENTUNO + CENTODUE +
       CENTOTRE + DUECENTO + TRECENTO = OTTOCENTO
```

```
(9 + 11 + 21 + 23 + 31 + 102 + 103 + 200 + 300 = 800)
```

```
8246 + 983010 + 4687982 + 46870756 + 75687982 + 16872396 +

 16872756 + 39616872 + 75616872 = 277216872
```

2842. 800 in Italiano – Due by Giulio Cesare, Rome, Italy

```
SETTE + OTTO + NOVE + TRENTUNO + TRENTOTTO + NOVANTA +

 CENTO + CENTOOTTO + CENTONOVE + TRECENTO = OTTOCENTO

(7 + 8 + 9 + 31 + 38 + 90 + 100 + 108 + 109 + 300 = 800)
```

2843. Four Month Sentence by Andrzej Bartz, Fuerth, Germany

```
    MAY + JUNE + JULY + JANUARY

   650 + 1278 + 1230 = 1572590
```

2844. So Seuss Me! by Andrzej Bartz, Fuerth, Germany

$$(THE + CAT)^2 + (IN)^2 = (THE)^2 + (HAT)^2$$

$$(386 + 513)^2 + (42)^2 = (386)^2 + (813)^2$$

2845. Twenty-two Countries – 1 by Frank Mrazik, Montreal, Quebec

Solve in base 14:

```
ALBANIA + ANDORRA + ARUBA + ASCENSION + BAHRAIN + BARBADOS +

BARBUDA + BELARUS + BOSNIA + BRUNEI + BURUNDI + CANADA +

ECUADOR + HONDURAS + ICELAND + INDONESIA + ISRAEL +

LEBANON + LIBERIA + RUSSIA + SERBIA = SEYCHELLES

a9ba74a + a76088a + a83ba + a15271407 + bad8a47 +

ba8ba601 + ba8b36a + b29a831 + b0174a + b83724 +

b383764 + 5a7a6a + 253a608 + d07638a1 + 4529a76 +

47607214a + 418a29 + 92ba707 + 94b284a + 83114a +

128b4a = 12c5d29921
```

2846. Twenty-two Countries – 2 by Frank Mrazik, Montreal, Quebec

Solve in base 15:

ALBANIA + ANDORRA + ARUBA + ASCENSION + BAHRAIN + BARBADOS +

BARBUDA + BELARUS + BOSNIA + BRUNEI + BURUNDI + CANADA +

ECUADOR + HONDURAS + ICELAND + INDONESIA + ISRAEL +

LEBANON + LIBERIA + RUSSIA + SERBIA = SEYCHELLES

9e79589 + 952bdd9 + 9b679 + 91c3518d5 + 794b985 +

79b792d1 + 79b7629 + 73e9b61 + 7d1589 + 7b6538 +

76b6528 + c95929 + 3c692db + 4d526b91 + 8c3e952 +

852d53189 + 81b93e + e3795d5 + e873b89 + b61189 +

13b789 = 13ac433331

2487. Skeleton – F by Junya Take, Kanagawa, Japan

(All F's are shown, * ≠ F)

```
        * * * * * *
        * * * * * *
        _____
        * * * * * *
        *FFF***
      ***F***
       ****FF*
    ******F
    _____
  ******F******
```

```
          1222088
          3057026
         ─────────
          7332528
         2444176
        8554616
       6110440
      3666264
    ─────────────
    3735954790288
```

2848. Skeleton – E by Junya Take, Kanagawa, Japan

(All E's are shown, * ≠ E)

```
        * * * * * * *
        * * * * * * *
       ─────────────
         * * * * * * *
        * * * * * * *
       * * E E E * * *
      * * * * E * * *
     * * * * * * E E *
    * * * * * * * E
   ────────────────────
   * * * * * * * E E E * * * * *
```

```
          11222383
          80960251
         ──────────
          11222383
          56111915
         22444766
        67334298
       101001447
      89779064
     ──────────────
     908566944498133
```

PROJECT EULER

URL: https://projecteuler.net

Charles Ashbacher

Project Euler is an online collection of problems that generally require a combination of mathematical insight as well as computer programming skills. The intended audience consists of people who are looking for challenging problems in mathematics that will stretch their problem solving abilities and force them to simultaneously think mathematically and programmatically.

Participation is open to all people; all you need to do to get started is create an account so your success can be properly logged. Charts of the problems you have solved are available and there are virtual awards given after you have solved a certain number of problems. Level advancements are given for every 25 problems you solve, and you can solve them in arbitrary order.

As of the time of this writing there are 501 problems on the site, with a wide variety of difficulty. Some could serve as projects in fairly low-level programming classes. For example, problem one:

Multiples of 3 and 5

If we list all the natural numbers below 10 that are multiples of 3 or 5, we get 3, 5, 6 and 9. The sum of these multiples is 23.

Find the sum of all the multiples of 3 or 5 below 1000.

Finding this solution is well within the skill set of a first semester programming student.

Significantly more difficult is problem 250:

Problem 250

Find the number of non-empty subsets of $\{1^1, 2^2, 3^3,..., 250250^{250250}\}$, the sum of whose elements is divisible by 250. Enter the rightmost 16 digits as your answer.

Any person attempting to solve this by simply writing a Java program using the BigInteger class will quickly discover there is a combinatorial explosion.

Problem 500 is almost pure number theory:

Problem 500

The number of divisors of 120 is 16.
In fact 120 is the smallest number having 16 divisors.

Find the smallest number with 2^{500500} divisors.
Give your answer modulo 500500507.

Clearly, the solution requires the use of the number theory function that gives the number of divisors an integer has.

There is a statistics section so that you can easily compare your performance against members around the world. As I am writing this only 40 members have solved more than 475 problems. If you are looking for a way to challenge your math and computer skills, keep your mind sharp and compete with others, this is the place to look

New problems are generated by members and the site is funded by the membership.

MAP COLORING WITH COMBINATORICS

by Kate Jones

Ever since I came across pentominoes in 1976, I've been fascinated by the concept of puzzle sets that are made up of exactly one each of every possible combination and permutation of their unit building block (polyform puzzles). From there I branched out to other kinds of tilings with different shapes that fit together without holes.

Within a few years of starting the company, Kadon Enterprises, Inc., to produce classic and original combinatorial puzzles, we had researched and published over a dozen different sets. One of my design objectives was to differentiate certain pieces of any one set by color. For example, with different sizes of tiles, each size would get its own color. One of the challenges for these sets was to solve them with "map coloring": so that no two pieces of the same color were edge-wise connected. Vertices could touch.

In standard map coloring, not more than 4 colors are needed so no two adjacent spaces share a color. Here are a few examples using 4 colors and color separation. All product names are proprietary trademarks of Kadon:

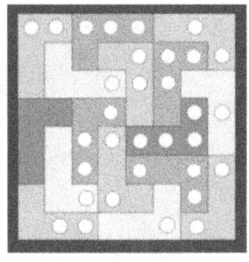 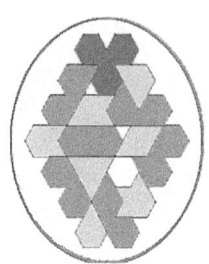 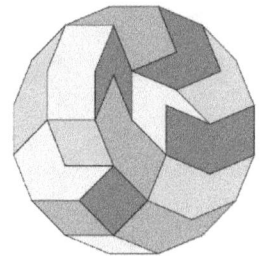 

L-Sixteen	**Star-Hex II**	**Rombix**	**Mini-Iamond Ring**
16 L-shaped tiles, each a different hole pattern, fill an 8x8 square, all holes linked	17 tiles of 1 to 3 hexagons and 1 to 3 triangles joined fill a star grid	16 rhombic tiles as singles and twins fill a 16-sided arena in over 19,000 ways	Even though this set has 15,140 solutions, only 2 solutions fit into the tray with full color separation.

Pentarose (Penrose prototiles)	**Leaves**
36 pentagon-related tiles with fractal edges fill a decagon tray	13 hexagon tiles with in and out arcs form a symmetrical star

To make the puzzles even more challenging, we wanted to be able to do map coloring with just 3 colors. Here are a few samples of these more exotic and artful pieces:

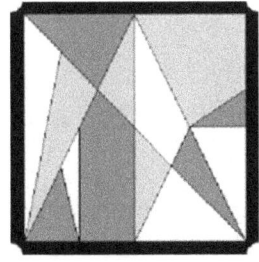

Trio in a Tray	**Archimedes' Square**	**Tangramion**	**Iamond Hex**
A special combo of 3 four-piece "Tiny Tans" brain	World's oldest puzzle, with 536 solutions. Not all separate the colors. The 3 colors divided among 14 pieces also have equal area.	7-piece lovechild of tangrams and Stomachion. Only 10 solutions in	12 hexiamonds with 4 of each color. Unique olutions for colors

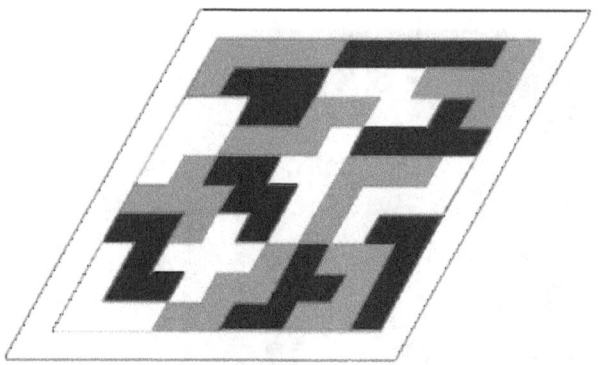

Rhombiominoes

20 pentarhombs in unique separation
mode. Unknown: how many will join all
colors. New solutions win prize.

Snowflake Super Square

36 tessellated squares (24 distinct)
have hundreds of goals. Unknown:
how many non-match solutions exist.

It is interesting to note that all these sets can also be solved so that pieces of the same color are connected into a single region—the min and the max. Between those endpoints are countless combinations for building other forms and figures. The open-endedness of each set makes for a lifetime of enjoyment. Each new solution brings surprise and joy, qualities that are difficult to measure mathematically.

These puzzles are eloquent paradigms of how people think, solve, design, and organize systems. Out of diversity comes harmony. Most are suitable for ages 12 to adult. You can see these and more in the full website at **www.gamepuzzles.com**

Mathematical Spectrum

A magazine for students
and teachers of mathematics
in schools, colleges and universities

Editor: D. W. Sharpe, *University of Sheffield*

For over three decades, *Mathematical Spectrum* has been a popular source of stimulating ideas for teachers, students and mathematical enthusiasts alike. Articles cover a wide range of topics in mathematics and the related sciences as well as the history of mathematics, with regular education and computer columns, a letters page, problems and solutions, and reviews of books and software.

Contributors from all over the world include established mathematicians as well as students — we welcome original student contributions and award annual prizes for the best ones published.

Subscription information: Vol. 47 (Sept 14–Aug 15) $24.50; Vols 47 and 48 (Sept 14–Aug 16) $47.00; Vols 47, 48 and 49 (Sept 14–Aug 17) $67.50. Three issues per volume in September, January and May; postage and handling included. To subscribe, contact:

Mathematical Spectrum	Tel: +44 114 222 3922
Applied Probability Trust	Fax: +44 114 222 3926
School of Mathematics and Statistics	
The University of Sheffield	Email: s.c.boyles@shef.ac.uk
Sheffield S3 7RH, UK	Web: www.appliedprobability.org

Published by the **Applied Probability Trust**, a non-profit-making organisation based in the University of Sheffield

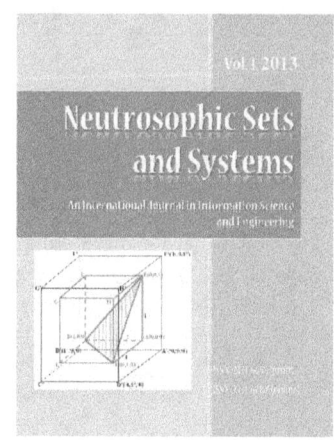

Editor-in-Chief:

Prof. Florentin Smarandache

Department of Mathematics and Science

University of New Mexico

705 Gurley Avenue

Gallup, NM 87301, USA

E-mail: smarand@unm.edu

Home page: http://fs.gallup.unm.edu/NSS

Associate Editors:

Dmitri Rabounski and Larissa Borissova, independent researchers.

Said Broumi, Univ. of Hassan II Mohammedia, Casablanca, Morocco.

A. A. Salama, Faculty of Science, Port Said University, Egypt.

Yanhui Guo, School of Science, St. Thomas University, Miami, USA.

Francisco Gallego Lupiañez, Universidad Complutense, Madrid, Spain.

Peide Liu, Shandong Universituy of Finance and Economics, China.

Pabitra Kumar Maji, Math Department, K. N. University, WB, India.

S. A. Albolwi, King Abdulaziz Univ., Jeddah, Saudi Arabia.

Mohamed Eisa, Dept. of Computer Science, Port Said Univ., Egypt.

Neutrosophic Sets and Systems has been created for publications on advanced studies in neutrosophy, neutrosophic set, neutrosophic logic, neutrosophic probability, neutrosophic statistics that started in 1995 and their applications in any field, such as the neutrosophic structures developed in algebra, geometry, topology, etc.

The submitted papers should be professional, in good English, containing a brief review of a problem and obtained results. Neutrosophy is a new branch of philosophy that studies the origin, nature, and scope of neutralities, as well as their interactions with different ideational spectra.

This theory considers every notion or idea <A> together with its opposite or negation <antiA> and with their spectrum of neutralities <neutA> in between them (i.e. notions or ideas supporting neither <A> nor <antiA>). The <neutA> and <antiA> ideas together are referred to as <nonA>.

Neutrosophic Set and Logic are generalizations of the fuzzy set and respectively fuzzy logic (especially of intuitionistic fuzzy set and respectively intuitionistic fuzzy logic). In neutrosophic logic a proposition has a degree of truth (T), a degree of indeter minacy (I), and a degree of falsity (F), where T, I, F are standard or non-standard subsets of $]^-0, 1^+[$.

Neutrosophic Probability is a generalization of the classical probability and imprecise probability.

Neutrosophic Statistics is a generalization of the classical statistics.

What distinguishes the neutrosophics from other fields is the <neutA>, which means neither <A> nor <antiA>.

<neutA>, which of course depends on <A>, can be indeterminacy, neutrality, tie game,

unknown, contradiction, ignorance, imprecision, etc.
All submissions should be designed in MS Word format using our template file:

http://fs.gallup.unm.edu/NSS/NSS-paper-template.doc

A variety of scientific books in many languages can be downloaded freely from the Digital Library of Science:

http://fs.gallup.unm.edu/eBooks-otherformats.htm

To submit a paper, mail the file to the Editor-in-Chief. To order printed issues, contact the Editor-in-Chief. This journal is non-commercial, academic edition. It is printed from private donations.

Information about the neutrosophics you get from the UNM website:
http://fs.gallup.unm.edu/neutrosophy.htm
The home page of the journal is accessed on
http://fs.gallup.unm.edu/NSS

Topics in Recreational Mathematics

Presenting papers and articles in recreational mathematics or material of interest to people interested in recreational mathematics. Original artwork with a mathematical theme will also be featured.

Contents

Editor-in-chief
Charles Ashbacher

Assistant editor
Rachel Pollari

Artwork
Caytie Ribble

Technical assistant
Gisela Hausmann

Dedicated to the legacy of Martin Gardner and Joseph S. Madachy

Available on Amazon ISBN 978-1507603215

ALPHAMETICS AS EXPRESSED IN RECREATIONAL MATHEMATICS MAGAZINE

Alphametics have been a staple of recreational mathematics since the first issue of **Recreational Mathematics Magazine**. A column of alphametics appeared in the first issue of RMM and it was a regular feature in Journal of Recreational Mathematics throughout the 38 ½ volumes that it was published.

This book contains the alphametics and their solutions that appeared in Recreational Mathematics Magazine during the 14 issues that it was published by Joseph S. Madachy.

Contents

Editor's Notes
by Charles Ashbacher

Mathematical Cartoon
by Caytie Ribble

Introduction
by Charles Ashbacher

Solving Addition Alphametics
by Charles Ashbacher

The Alphametics That Appeared in Recreational Mathematics Magazine

Solutions to Alphametics

Available on Amazon

ISBN 978-1508538134

Editor-in-chief
Charles Ashbacher

Artwork
Caytie Ribble

Technical assistant
Gisela Hausmann

www.ingramcontent.com/pod-product-compliance
Lightning Source LLC
Chambersburg PA
CBHW080642180526
45168CB00008B/3278